Key Elements of Food Safety and Food Processing

KEY ELEMENTS OF FOOD SAFETY AND FOOD PROCESSING

Edited by:

Reema

Delve
PUBLISHING

www.delvepublishing.com

Key Elements of Food Safety and Food Processing

Reema

Delve Publishing
224 Shoreacres Road
Burlington, ON L7L 2H2
Canada
www.delvepublishing.com

Email: orders@arclereducation.com

ISBN: 978-1-77469-475-6 (Hardcover)

ABOUT THE EDITOR

Dr. (Mrs.) Reema is presently working as Guest Faculty, Department of Home Science, Government Degree College, Katra Chuggupur, Sultanpur, Uttar Pradesh. She graduate from the Deen Dayal Upadhyay Gorakhpur University and completed doctorate from Nehru Gram Bharti Deemed to be University, Prayagraj in the year 2020 with UGC NET, JRF &SRF. She has published four research paper in elite national publications and participated many national & International conferences.

TABLE OF CONTENTS

LIST OF FIGURES

ABSTRACT

Food poisoning is among the most widely recognized reasons for sickness around the world. This volume seeks to educate readers on food safety and gives balanced arguments on pesticides, and genome modification among other elements influencing food safety. Billions of individuals on the planet are in danger of unsafe food. A large number become sick while others die from eating unsafe food. Thusly, safe food saves lives. Safe food enriches the individual and population health. Safe food further creates economic development of the location where sanitation is improved. Safe food supply relies upon both science and effective food policing. With technological improvements, new guidelines should be ratified to create a regular supply of food items that are protected and fit for consumption.

As living standards improve, worries over food safety and possible impurities will continue being a significant medical concern. Consumers request quality and safe products they eat since food provides the necessary energy and nutrients for supporting life. Largely, consumers depend on the government to protect food items and guarantee sold foods meet minimum consumer safety standards. For instance, a container of olive oil named as 100 percent virgin olive oil should contain the very thing the name says aside from the naturally occurring minor components that are essential for the olive oil to last, and which can't be removed or disposed of totally without affecting the olive oil. The volume examines challenges and strategies in food safety and handling across different countries worldwide.

PREFACE

Food can never be completely protected, however, authorities can try to make the situation better for consumers. Safety is undermined by various microorganisms that cause diverse foodborne infections, algal poisoning that cause severe illness, and fungal poisoning that might be deeply harmful may likewise have terminal sequelae, for example, immunotoxic, nephrotoxic,teratogenic, and estrogenic impacts. Maybe more concerning, the modern advances of the last century have brought about enormous expansions in our exposure to harmful metals like mercury, lead, cadmium, and arsenic, which show different poison levels. This volume mentions that modern development additionally brought synthetic substances that, albeit prohibited in the past, persevere in the climate and contaminate our food.

These comprise organochlorine compounds, for example, (dichloro diphenyl dichloroethene) (DDT), different pesticides, dioxins, ethane and dioxin-like mixtures. DDT and its constituents such as dichlorophenyl dichloroethylene affect the male and female reproductive organs. Also, there is growing proof that they cause neurodevelopmental poison levels in human babies and youngsters. They share this trademark with the dioxins. Other food toxins can emerge from the treatment of animals with veterinary medications or the showering of food crops, which might leave buildups. Among the pesticides applied to food crops, such as organophosphates have been the focal point of much regulatory consideration since there is developing proof that they, as well, affect the developing mind. Various compound pollutants are formed during the handling and cooking of food varieties. Large numbers of them are known or thought to be cancer-causing agents. Other food toxins filter from the storage or stockpiling chambers. Examples that stand out enough to be noticed are phthalates, which have been shown to cause deformities in the male reproductive system in research, and bisphenol A, which adversely affects the development of the central nervous system and the male organs.

The volume further argues that genetically altered food sources present new challenges to authorities, since customers fear that the conceivable health risks of these food varieties have not been explained. An arising danger to safety conceivably comes from the growing utilization of nanomaterials, which are now added to storage materials, despite the fact that their harmfulness remains generally neglected. Various groups have highlighted the significance of resolving this issue and creating the necessary instruments for addressing these issues. Legislative centers, for example, the US Food and Drug Administration and their partners in different countries have undeniably faced challenges in checking food supplies for these synthetic compounds and deciding the human health hazards related to exposure to these substances.

The methodology taken up to this point zeroed in on each compound in turns. It is progressively observed, nonetheless, that a considerable lot of the different synthetics we are exposed to regularly are prevalent, coming from contaminated water, air, residue, and soil. "Mixture toxicology" is a quickly developing science that tends to address connections among synthetics and explore the impacts of aggregate exposure to these "regular mechanism groups" of synthetic substances. It is to be believed that this results in a more profound comprehension of the risks faced from various common exposures, and contributes to making our food supplies more secure.

CHAPTER 1

IMPORTANCE OF FOOD SAFETY

CONTENTS

1.1 INTRODUCTION

Food safety pertains to procedures for preparing, handling, and storing food that are designed to avoid foodborne sickness and harm. During their journey from farm to the manufacturer to fork, food products may come into contact with a variety of health dangers (Zanin et al., 2017). To reduce these hazards and protect customers, safe food handling methods and procedures are applied at every phase of the food production process. Food safety is a scientific subject that depends on a variety of academic disciplines, including engineering, microbiology, and chemistry. These several schools of thought come together to assure food processing safety wherever food is obtained, made, processed, stored, or marketed. Food safety, in this view, is a holistic approach to cleanliness and responsibility that encompasses all aspects of the international food sector.

Figure 1.1: Food safety refers to the lack, or safe, accepted levels, of dangers in food that might impair customers' health. It refers to reducing the presence of dangers, whether prolonged or acute, that may cause food to be harmful to the consumer's health.

Source: https://www.publichealthnotes.com/food-safety-what-you-should-know/

Food is one of the most traded products on the planet. With each New Year, markets become more globalized, and the world's demographic continues to expand, the international food supply chain would only grow in scale and complexity. Food safety adherence has never been more critical than it is now, because of these megatrends driving the mass distribution and production of food. Every nation has its own regulatory agencies in charge of defining and enforcing local food safety regulations. Domestic and foreign enterprises alike are liable to the food safety regulations and enforcement mechanisms of any particular country in order to produce or sell food commodities (Xiang et al., 2019).

Many privately-owned international institutions also offer extensive recommendations for evaluating food businesses on food safety and cleanliness. These international standards enable food industry stakeholders from many nations to guarantee that food quality and safety requirements are met in a cross-border manner, facilitating global food commerce. Global market giants in the food sector frequently pursue certification with a variety of private food regulators in addition to adhering to the food safety legislation of the countries in which they operate. They may also seek verification of the same certifications from the downstream and upstream suppliers with whom they operate.

International regulatory agencies provide auditing, certification, and guidance services to global food businesses, but they are not in charge of enforcing food safety rules. Every country defines and sets its own food safety rules and enforcement techniques, which may differ from country to country and domestically from area to region (Ambrus, 2016). Introducing a food item to a foreign market necessitates adherence to that country's food safety and consumer protection rules, as well as those of its regional governments. In general, international food safety standards are intended to make it easier to comply with food safety legislation in large markets by streamlining the clearance process with foreign government authorities.

Since the beginning of time, foodborne diseases have posed a hazard to human health. Many of the food preparation processes we use today, such as heating, canning, smoking, and fermentation, may be thought of as primitive food safety procedures created to avoid people from becoming ill. Today, we enjoy the fruits of decades of scientific and technical advancement, which have resulted in a variety of safe food and drink items that many of us take for granted. However, the modern idea of food safety, as well as the rigor with which it is implemented, is a comparatively recent development in human history, closely linked to changes in how we live and eat.

Upton Sinclair, an American novelist, wrote his novel The Jungle in 1905, which included graphic portrayals of Chicago's meatpacking industry. After the public outcry, the US government passed the Meat Inspection Act the next year, creating the first hygienic slaughtering and butchering requirements. This law was one of the first to subject food processing plants to regular government audits and inspections, as well as one of the first regulations governing food safety in production.

Figure 1.2: Food safety is a critical component of the food business, and it is impossible to overestimate its importance. The food chain, whether you're a producer, a customer, or a distributor, keeps everyone in the process controlled. Food globalization may have brought food from all over the world to your doorstep, but it also puts customers in danger of food poisoning. That is why the topic of food safety is brought up in the first place.

Source: https://www.trainingexpress.org.uk/what-is-food-safety/

The industrial revolution brought in the formation of several regulatory agencies and basic legislation governing food safety and inspections across Europe and North America. Laws were created to prevent the purposeful selling of food items that were misbranded, polluted, or otherwise interfered with as food production grew increasingly mechanized and economic incentives increased. Ingredients and additions began to be regulated during this time period. Electric refrigerators reached middle-class homes all over North America and Europe in the decades after WWII, revolutionizing the way people bought and kept food. Home refrigeration ushered in a new age of industrial production of food, as well as a rising demand for more stringent food controls. In 1947, Mars Incorporated was the first large food company to use metal detectors in their plants, owing to the shifting food scene (Wilcock et al., 2004).

When HACCP was founded in 1959, it marked the beginning of a transition from reactive to proactive food safety standards. Recognizing that the testing of finished products was ineffective in ensuring food safety and quality, NASA scientists teamed up with the Pillsbury Company, an American bakeware and baking mix manufacturer, to develop a risk-based system that recognized "critical failure areas" in manufacturing that presented health risks. This hazard analysis and control approach was implemented by a number of large food producers in the United States, with Pillsbury taking charge.

By the mid-1980s, experts all around the globe had concluded that HACCP was a more effective way of addressing food safety concerns than traditional inspection techniques due to its proactive character. In an increasingly globalized world food business, international regulatory agencies and third-party audit firms were established in the following decades to adopt and enforce preventive compliance. Modern food safety standards and practices are founded on this foundation.

1.2 FOOD SAFETY AS A GLOBAL HEALTH ISSUE

Food safety is critical in order to safeguard customers from common allergies and foodborne diseases. This is a good enough aim to achieve utilizing correct processing and handling processes on its own, but there are additional reasons as well. Food safety products protect businesses and stakeholders from costly fines and legal action (Wertheim-Heck et al., 2019). Fines and legal ramifications might force a plant to shut or even ruin a firm. By offering training and instruction to everyone who handles ingredients in any food industry, facilities may accomplish adequate food safety procedures. Following these precautions meticulously is critical for protecting your consumers from foodborne illness, allergic reactions, and other health hazards associated with tainted food.

The production of safe food is influenced by a number of elements that cover the whole process, from harvesting to processing to packing. Agricultural techniques, labor practices, the application of preventative measures during food processing and preparation, the use of chemical materials, the proximity of raw ingredients and water, and storage are all examples. Hygiene is a crucial component for optimal standards at every level. Poor preservation, storing cooked and raw meals together, and employing inappropriate procedures to prepare, heat, and store components can all compromise the product's sanitary quality.

Many public health agencies underestimate the worldwide relevance of food safety. The frequency of foodborne disease has been steadily increasing, according to epidemiological surveillance. Furthermore, both affluent and developing nations have had severe epidemics of salmonellosis, cholera, enterohaemorrhagic Escherichia coli illnesses, hepatitis A, and other illnesses (Wahyuni et al., 2019). Cholera and other diarrheal illnesses are mostly foodborne, despite the fact that they are typically thought to be spread via water or person-to-person contact. Foodborne infections may affect up to 10% of the population in developed nations each year.

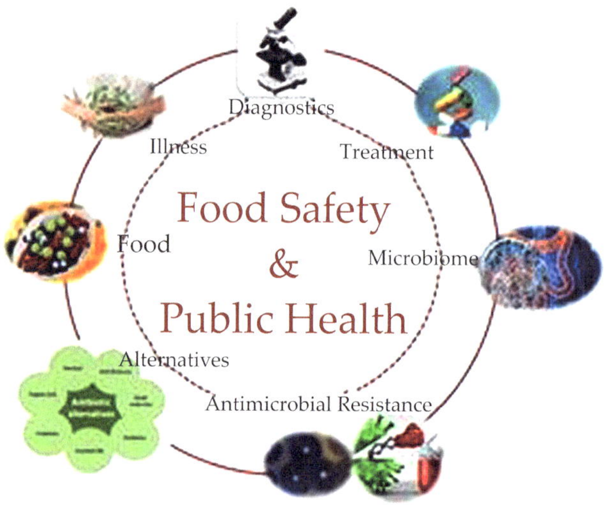

Figure 1.3: Foodborne disease has a significant influence on food safety and human health across the world. According to the Centers for Disease Control and Prevention (CDC), foodborne infections cause roughly 128,000 hospitalizations and 3,000 deaths in the United States each year. They cost the healthcare sector and the general public billions of dollars each year.

Source: https://www.frontiersin.org/research-topics/20666/food-safety-and-public-health

Transgenic foods, harmful chemicals in food, food irradiation, and the hypothetical risk of "mad cow" disease transmission through beef consumption have all piqued the public's curiosity. Food safety is anticipated to attract more attention in the twenty-first century, particularly as some worldwide developments, which are already underway, are predicted to have primarily negative consequences in this area. Among the reasons under concern include urbanization, changes in microbial as well as other ecological processes, and declining food and fresh water sources. However, a far more significant problem is anticipated in conjunction with changes that directly impact sanitation and the near human environment.

1.3 EVOLVING FACTORS IN IMPORTANCE OF FOOD SAFETY

1.3.1 Changes in Demography

Within two decades, the global population is estimated to exceed 8.5 billion people, with 80 percent of them living in developing nations. This is up from 5.8 billion in 1996. This massive rise and unequal distribution are likely to result in substantial food security and safety issues, destruction of the environment, large-scale movement from rural to urban regions, and from poor to affluent nations, as well as significant ecological changes. The number of individuals over 60 years old in industrialized countries is expected to climb from 17% presently to 25% by 2025 (Tirado et al., 2010). In emerging nations, a similar phenomenon is occurring. Such a shift is likely to result in severe socioeconomic issues as well as the appearance of a large number of individuals who are less resistant to diseases, especially foodborne illnesses.

1.3.2 Environmental Factors

Chemical and biological contamination of locations where food is produced, processed, and eaten greatly increases the risk of foodborne illness. Increased pollution is inevitable as a result of population expansion, uncontrolled movement from rural to urban regions, and the resulting slum construction. In such situations, particularly in poor nations, drinking water sources and sanitation systems are put under increased strain, increasing the danger of foodborne pathogens spreading.

Temperature has a considerable impact on the occurrence of foodborne illnesses and intoxications. In temperate countries with lengthy hot summers, significant increases in such illnesses have been documented. According to the United Nations Intergovernmental Panel on Climate Change, between

2030 and 2090, average temperatures will climb by around 1.1 and 3.1 degrees Celsius over 1995 levels, respectively. Because the relationships involved are complicated and multifaceted, the worldwide influence on foodborne illness and other areas of human health is unexpected. The occurrence of cholera and dysentery, on the other hand, has been linked to the marine phenomena known as El Nino. This emphasizes the importance of precise predictions of this and other events in order to take preventative actions against the illnesses in question.

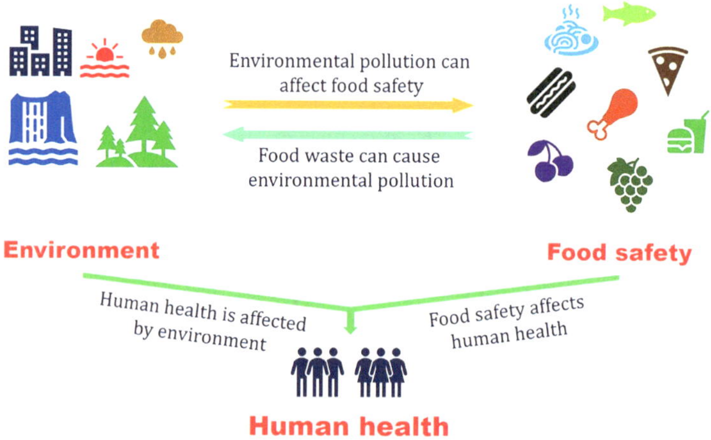

Figure 1.4: The environment and food safety have long been two of the world's most pressing issues. Human activities have put the environment and food safety in jeopardy. For example, huge amounts of garbage are created as a result of the widespread usage of disposable masks, which are mostly constructed of non-biodegradable polymers. In truth, the environment and food safety are inextricably linked.

Source: https://link.springer.com/article/10.1007/s11356-021-16069-6

Toxic substances discharged into the environment as a result of industrial activities and farming techniques may find their way into the human food supply. The health effects are expected to be negligible when little amounts are present in meals. Pesticides are known to inhibit the immune response in experimental animals, which is one reason for caution in this field. Pesticides have been detected in human tissues, particularly fat, in both developing and developed nations, but their impact on the body's immune system has yet to be fully investigated (Todt et al., 2009). Nonetheless, they are likely to inhibit the immune system in humans. Given the pressures of population growth, the usage of agrochemicals and the emission of industrial pollutants are projected to increase in the next decades. The effects might be disastrous, particularly for the 20 million youngsters in underdeveloped nations whose disease resistance has been weakened by starvation.

1.3.3 Behavioral and Social Factors

Poverty and unfairness are the most significant contributors to ill health. Poverty has been dubbed the world's most deadly disease. Because foodborne infections are common in both wealthy and poor nations, the divide between affluent and unprivileged people in food safety may appear less pronounced than in other areas. People in wealthy societies, on the other hand, typically suffer from mild diseases that remain as a result of risky lifestyles (preference for raw foods, food mishandling, etc.), whereas serious, life-threatening illnesses like infant diarrhea, cholera, typhoid fever, and fluke infection are still common in poor communities and cause high rates of mortality (Sinha, 2009). Approximately a fifth and a quarter of the world's populace lives in absolute poverty; this number is rising and will certainly continue to rise. Poverty is likely to be the most significant barrier to health-care equity, especially the prevention of foodborne infections.

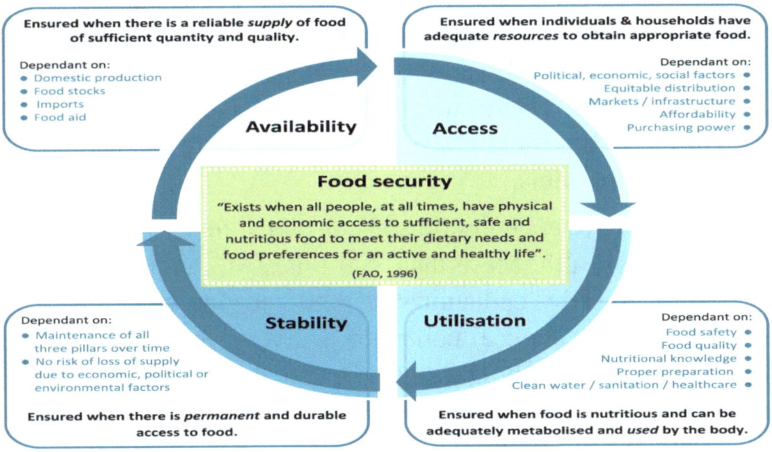

Figure 1.5: Food costs fluctuate owing to fluctuations in demand among metropolitan populations, creating a barrier to people's food intake. Political disputes may emerge as a result of a collision of ideas among a large population, causing food security to be overlooked.

Source: https://www.publichealthnotes.com/food-security-determinants-and-urbanization/

Foodborne illnesses are heavily influenced by one's behavior and lifestyle. In wealthy cultures, where customers seek minimally processed foods with lengthy shelf life, no preservatives, and little salt and sugar content, the dangerous practice of eating shellfish as well as other foods uncooked is becoming more widespread (Smith &Pillai, 2004). Pathogens are more prone to grow to harmful levels under such settings, even at refrigeration temperatures, increasing the risk of infection and poisoning. Because of the inherent virtues of the technologies and the initiatives of health educators, a consumer worries about food irradiation, an economical technique of making food safe, even in its raw condition, are anticipated to fade in the future century.

1.4 FOODBORNE DISEASES THAT ARE IMPORTANT IN FOOD SAFETY

Foodborne diseases and other food-related injuries should be avoided. The primary goal of food safety is to safeguard food consumers from foodborne infections or injuries caused by food consumption. As a result of poor food safety, foodborne infections are a huge danger to food industries and harm people all over the world. Foodborne pathogens, such as dangerous viruses, parasites, yeasts, fungi, and bacteria, are mostly responsible for these consequences. Foodborne diseases and acute poisoning can also be caused by chemical compounds, heavy metals, and excessive additives. A pathogen is likely to thrive in the absence of sufficient food safety controls, depending on the nature of your items.

It is critical to have enough safe and nutritional food to maintain life and promote excellent health. More than 200 illnesses, ranging from diarrhea to cancer, are caused by contaminated food carrying hazardous bacteria, viruses, parasites, or chemical compounds. Every year, an estimated 600 million individuals – nearly one out of every ten people on the planet – become unwell after eating tainted food, with 420 000 deaths, leading to the loss of 33 million healthy life years (DALYs). Each year, hazardous food costs low- and middle-income nations US$110 billion in productivity loss and medical costs (Azevedo et al., 2014).

Foodborne sickness affects 40% of children under the age of five, resulting in 125 000 fatalities each year. Diarrheal disease disorders are the most frequent illnesses caused by contaminated food, with 550 million people falling ill and 230 000 people dying each year. Food safety, nutrition, and security are all intertwined. Unsafe food feeds a vicious cycle of sickness and hunger that disproportionately affects newborns, children, the elderly, and the sick. Foodborne illnesses stymie socioeconomic growth by putting a load on health-care systems and negatively impacting country economies, trade, and tourism. Food supply systems increasingly span many countries. Governments, producers, and consumers working together to guarantee food safety are beneficial.

Safe food supplies are important for national economies, trade, and tourism, as well as food and nutrition security and long-term development. The number of individuals purchasing and consuming food produced in public areas has risen as a result of urbanization and shifts in consumer behavior, particularly travel. Consumer demand for a larger range of foods has increased as a result of globalization, resulting in a more complicated and longer global food chain. The industrialization and intensification of agricultural and animal production to fulfill rising food demand present both possibilities and difficulties for food safety as the world's population expands. Climate change is expected to have an influence on food safety.

Figure 1.6: According to the Centers for Disease Control and Prevention (CDC), around 48 million individuals (1 in 6) get sick from a foodborne disease each year, 128,000 are hospitalized, and 3,000 die. Food can be contaminated by a variety of disease-causing microorganisms (such as parasites, viruses, and bacteria), chemicals, and other substances, resulting in a variety of foodborne diseases.

Source: https://www.cdc.gov/ncezid/what-we-do/our-topics/foodborne-disease.html

Food producers and handlers bear a higher share of the duty for ensuring food safety as a result of these problems. Because of the speed and breadth of product distribution, local problems can swiftly escalate into international crises. In the last decade, serious foodborne illness outbreaks have happened on every continent, frequently exacerbated by worldwide commerce.

Unsafe food is a worldwide health concern that puts everyone at risk. Infants, small children, pregnant women, the elderly, and those suffering from an underlying ailment are all at risk. Every year, 220 million children get diarrhea, with 96 000 of them dying. Foodborne illness feeds a vicious cycle of diarrhea and malnutrition, jeopardizing the nutrient intake of the most vulnerable. The International Conference on Food Safety, which took place in Addis Ababa in February 2019, and the International Forum on Food Safety and Trade, which took place in Geneva in February 2019, both emphasized the significance of food safety in reaching the Sustainable Development Goals (Antle, 2001). Food safety should be a public health priority for governments, since they play a key role in setting laws and regulatory frameworks, as well as building and executing effective food safety systems.

Food can be contaminated at any moment during manufacturing and delivery, and food producers have the major responsibility. Foods incorrectly cooked or mistreated at home, in food service facilities, or at markets, however, generate a major proportion of foodborne illness occurrences. Not all food handlers and consumers are aware of their responsibilities, such as following fundamental sanitary standards while buying, selling, and preparing food in order to safeguard their own and the community's health.

Bacteria, chemicals, parasites, or virus compounds enter the body through polluted food or water and produce foodborne diseases, which are typically infectious or poisonous in nature. Foodborne bacteria can cause severe diarrhea or even life-threatening diseases like meningitis. Chemical pollution can cause immediate toxicity or long-term illnesses like cancer. Foodborne illnesses can cause long-term impairment and even death. Uncooked animal meals, feces-contaminated vegetables and fruits, and raw shellfish with marine biotoxins are all examples of dangerous foods. Some of these include:

1.4.1 Bacteria

Campylobacter, salmonella, and Enterohaemorrhagic *Escherichia coli* are only a few of the most frequent foodborne infections that afflict millions of people each year, sometimes fatally. Vomiting, nausea, headache, fever, stomach discomfort, and diarrhea are some of the symptoms. Poultry, eggs, and other animal-based products are examples of foods linked to salmonellosis epidemics (Baluka & Rumbeiha, 2016). Raw milk, raw or undercooked chicken, and drinking water are the most common sources of Campylobacter infections. Undercooked meat, unpasteurized milk, and fresh vegetables and fruits are all linked to enterohaemorrhagic *Escherichia coli*.

Figure 1.7: In chilly, damp conditions like the refrigerator, certain bacteria may survive and even proliferate. Fresh vegetables and fruits should be kept apart from raw poultry, meat, shellfish, and eggs in the refrigerator. To avoid cross-contamination, clean your refrigerator with hot water and soap on a regular basis, and clear up beverage and food spills right away.

Source: https://www.foodsafetynews.com/2016/09/how-may-do-you-know-top-10-food-safety-myths-and-facts/

Infection with Listeria causes miscarriage in pregnant women and the death of newborn newborns. Despite the disease's low prevalence, listeria has severe and often deadly health implications, especially in newborns, children, and the elderly, making it one of the most dangerous foodborne illnesses. Listeria may thrive at refrigerator temperatures and is found in unpasteurized dairy products and a variety of ready-to-eat foods.

People become infected with *Vibrio cholerae* after drinking contaminated water or eating contaminated food. Abdominal discomfort, vomiting, and a lot of watery diarrheas are all symptoms that can lead to severe dehydration and mortality. Cholera epidemics have been linked to millet gruel, vegetables, rice, and various types of seafood (Bech-Larsen & Aschemann-Witzel, 2012).

Antimicrobials, like antibiotics, are required to treat bacterial infections. Their abuse and misuse in human and veterinary medicine, on the other hand, have been related to the formation and spread of antimicrobial resistance bacteria, rendering infectious illness therapy in animals and people ineffective. Animals introduce resistant germs into the food chain (e.g., Salmonella through chickens). One of the most serious concerns of modern medicine is antimicrobial resistance.

1.4.2 Viruses

Watery diarrhea, nausea, vomiting, and stomach discomfort are all symptoms of norovirus infections. Hepatitis A is a virus that causes long-term liver illness and is spread by eating raw or undercooked fish or eating infected raw foods. Food contamination is frequently caused by infected food workers.

1.4.3 Parasites

Some parasites are only spread by food, such as fish-borne trematodes. Tapeworms like *Taeniasolium* or *Echinococcus* spp., for example, can infect people via foods or direct contact with infected animals. Other parasites that enter the food chain through soil or water, like Giardia, *Entamoebahistolytica*, Cryptosporidium, or Ascaris can infect fresh vegetables.

1.5 FINANCIAL COSTS OF FOOD SAFETY

Food safety dangers in our food that exceed specific pre-determined levels can result in significant losses for society as well as the agro-food business, which includes retailers, processors, and farmers. The economic effect of foodborne infections can be estimated from a variety of perspectives, such as from a societal or consumer perspective, and will include various cost components based on these factors (Bingham & Lavau, 2012).

1.5.1 Societal Costs

The societal costs of foodborne illnesses – the disease burden – might aid in prioritizing food safety concerns for the allocation of limited resources to prevention and management. Medical expenditures, productivity loss, quality of life, and death are all costs to society. Disease burden is commonly evaluated in terms of Disability Adjusted Life Years (DALYs) or, less frequently, Quality Adjusted Life Years (QALYs) or cost-of-illness (CoI). DALYs are the total number of years of life lost owing to premature death plus the number of years spent disabled. A DALY is a unit of measurement that reflects one year of healthy life lost (Boudergue et al., 2009).

The cost of a foodborne disease in a specific timeframe is estimated using the CoI approach. Direct health expenses (e.g., hospitalization, medications), direct non-health expenditures (e.g., transportation, informal care), and indirect non-health costs (e.g., lack of availability at work) are all considered in this strategy. Because the CoI approach necessitates a large amount of data on patient medical procedures, it is not commonly used. Furthermore, CoI estimates have been criticized for failing to account for all costs incurred by foodborne diseases since they never include a monetary value for patients' pain and suffering.

Figure 1.8: Rather than being proactive, most responses are reactive - in response to large foodborne illness outbreaks or trade disruptions. Countries may improve their farmers' and food industry's competitiveness and build their human resources by concentrating more intently on domestic food safety.

Source: https://www.foodsafetynews.com/2018/10/unsafe-food-in-lmics-costs-110-billion-a-year-world-bank/

The Foodborne Disease Burden Epidemiology Reference Group (FERG) was established by the World Health Organization (WHO) in 2007 to approximate the highest prevalence of foodborne diseases – in terms of DALYs – attempting to cover acute effects of microbial contamination as well as acute and long - term effects associated with the presence of dangerous substances in food. The study's overarching purpose was to compile existing data and support national research on illness burden. Until this WHO study, studies on the burden of illness had solely concentrated on microbiological risks; for the first time, chemical hazards such as aflatoxins and dioxins were included in the burden of disease estimations for the first time.

1.5.2 Costs in the Agro-food industry

In the agro-food sector, food safety management is an integral aspect of food quality management. The existence of food safety hazards in doses above pre-determined limits can result in costs for the agro-food industry, such as costs to comply with regulations, costs to trace back the contamination, product defects, plant closure and cleaning, liability insurance, and the market's long-term effect due to reputational harm. The majority of research focused on estimating the direct costs of food safety issues in the United States, such as the expenses of informing customers, recalls, and litigation (Sargeant et al., 2009). Due to the scarcity of data needed to determine the costs of food safety accidents, some writers created a hypothetical incident, concentrating on the approach and modeling the possible economic effects for various stakeholders throughout the supply chain.

These models enable the analysis of changes in economic losses when food safety preventative and control methods, as well as enhanced monitoring, are implemented. These cost estimates demonstrate the value of increased food safety to businesses, as well as the costs and advantages of traceability. Food safety incident costs can be used to create a budget for preventative measures and choose the most cost-effective choices.

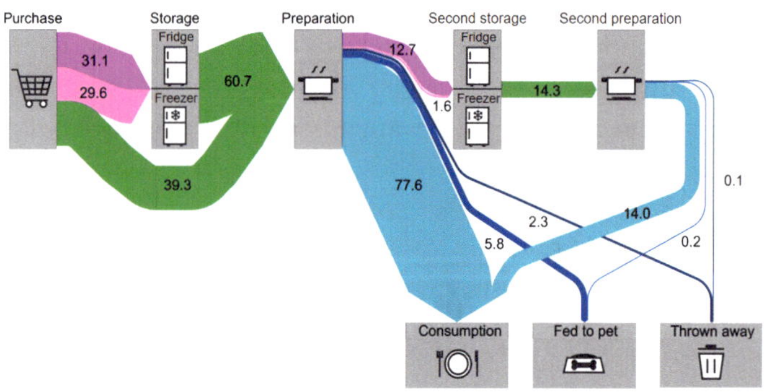

Figure 1.9: A third of the food manufactured for human use is lost or squandered. Significant progress has been made in quantifying food waste volumes by area and commodity category, as well as the environmental effect of that waste.

Source: https://www.mdpi.com/2071-1050/13/13/7027

The food business in the United States has a $55.5 billion food safety concern. Foodborne infections in restaurants, motels, convenience shops, and other food service establishments, as well as food recalls and other food safety problems, are all covered. Clearly, restaurants should be worried about sickening—or perhaps killing—customers as a result of a foodborne disease, and should take all reasonable precautions to minimize that risk. Foodborne infections, on the other hand, are costly and destructive to enterprises. A foodborne disease outbreak may cost restaurants a lot of money, including lost income, high legal bills, prospective litigation, decreased sales (and loyalty) from anxious customers, and a tarnished reputation that could force them to close their doors permanently.

According to research by Ohio State University professor Robert Scharff, foodborne diseases cost the United States $55.5 billion per year in medical care, productivity loss, and illness-related death. As per a new study from scientists at Johns Hopkins Bloomberg School of Public Health, the cost of a single foodborne illness epidemic at a fast-casual institution could range from $6,330 to $2.1 million, based on the intensity of the outbreak, the proportion of lawsuits, fines, and legal fees, and also the number of guests and employees affected by the incident. According to the report, casual-dining restaurants might incur expenditures ranging from $8,030 to $2.2 million each incident, while fine-dining restaurants could incur costs ranging from $8,273 to $2.6 million per incident (Borchers et al., 2010).

According to the Centers for Disease Control and Prevention, 48 million Americans get sick from foodborne diseases each year. In the United States, a 1% reduction in foodborne disease would prevent nearly 500,000 individuals from becoming ill each year. Five million people would be saved if foodborne diseases were reduced by 10%. One of the best things restaurants (and other food service enterprises) can do is invest in food safety. The cost, time, and effort required to implement—or improve—your food safety practices will not be prohibitive, and it is critical to your company's success.

Figure 1.10: All vegetables and fruits should be properly rinsed. Before preparing them, immediately rinse them with running water. Scrub hard fruit like cucumbers or carrots using a vegetable brush under running water. Spills should be cleaned up as quickly as possible, and every surface that comes into touch with food should be cleaned routinely. Clean the interior and exterior of appliances, giving specific attention to knobs, handles, and buttons where cross-contamination might occur due to hand contact.

Source: https://www.consumernotice.org/products/food-safety-guide/

Reduce the risks of foodborne disease episodes and outbreaks, as well as the enormous expenses connected with them, by implementing the following steps. Make food safety a company-wide priority. Food safety rules must be taught to everyone on every shift. Invest in cutting-edge technology. Many restaurants are using digital technologies to improve their operations. These technology technologies not only make food safe, but they may also save businesses a lot of money by minimizing food waste and lowering the risk of foodborne illness. Get rid of the paper. Technical solutions improve food safety processes by making inventory, monitoring, training, and keeping food safe faster, more precise, and more efficient. Restaurants that accept and embrace digital solutions can assist improve their operations' health and safety (Bolek, 2020).

Reduce the possibility of human mistakes. While human error will never be totally eradicated, technological developments can assist to reduce the dangers. Sensors guarantee that meals are kept at the correct temperature. When temperatures in restaurant refrigerators or freezers climb over safe holding temperatures, integrated, continuous cooling monitoring systems alert operators, removing the need to throw away whole coolers or freezers of food due to malfunctioning machines. As a consequence, restaurants may save tens of thousands of dollars (or more) in wasted goods while also possibly saving lives.

Improve the quality of your data collecting. Restaurants may now employ innovative digital technologies for internal auditing processes, which is a more efficient, cost-effective, and accurate option than the pen and paper techniques that are commonly used in the business. Auditing restaurants using a pen and paper generally results in more work, time, mistakes, and costs. It's tough to manage and retrieve hard copy records, especially in the middle of a terrifying food safety epidemic, and it's much more difficult to integrate and analyze the data. Internal auditing systems that use digital technologies are more efficient and cost-effective, with records that are easier to access and analyze (Redmond & Griffith, 2004).

Make yourself mobile. Millennials (or younger) make up the majority of restaurant personnel, and they are addicted to their phones. There's a huge contradiction between emphasizing the necessity of food safety measures and providing personnel with outmoded pencil and paper record-keeping tools. Instead, put in place communications technology that can be tracked via smartphones and tablets. Employees may use downloaded applications to improve the way they perform inspections, record temperature logs, train, handle QA forms, and access food code information, among other things. This puts essential food safety information (literally) at the fingertips of staff. Increase the efficiency of your operations. You will increase your restaurant's bottom line by doing so. According to a recent CoInspect survey, eliminating pen and paper line checks may save $250-600 per year for each restaurant, smart sensors that prevent food deterioration can save $1,100 every episode, and lowering food and labor expenses can save $4,700 per year per restaurant. By streamlining and enhancing line checks, shift records, inspections, audits, and other reporting, digital solutions may aid with brand protection and quality assurance problems (Rendueles et al., 2011).

Reduce the amount of pencil whipping. In the food service business, there is a widespread "pencil whipping" problem, in which employees who use paper record systems falsify records or "cheat" on their processes. As much as restaurants, lodgings, corner stores, and other industry enterprises would like to deny that "pencil whipping" occurs at their establishments, it is (sadly) a pretty widespread practice. Increased food safety concerns, food code violations, and other (possibly costly) difficulties can all occur from pencil whipping. Through real-time data collecting and visual recordings utilizing images and videos, digital solutions can assist decrease or eliminate "pencil whipping."

While technology was formerly considered a luxury, it is now inexpensive, widely available, and easily accessible. Technology that may assist cut labor costs, reduce (or eliminate) foodborne disease risks, and reduce food waste is a significant investment, not an expenditure. To keep goods, consumers, and restaurants safe and healthy, innovative digital systems and records are essential. Restaurants must make food safety a top priority, employing the most efficient and effective technologies to safeguard the health of their customers, staff, and companies. Technology helps to simplify processes, enhance safety measures, decrease mistakes, integrate data, and much more, with huge advantages.

1.6 FOOD WASTAGE AND FOOD SAFETY

Processed foods that have been found to have food safety problems may be discarded. A food firm can reduce wastes as a consequence of a better food manufacturing process by employing effective food safety measures. Food safety management systems are also meant to detect and mitigate food risks before they reach the manufacturing process, which might result in additional waste if processed further. Food waste is another important issue. Food waste is a big problem for our natural resources, the environment, and our wallets. Food production in the United States consumes 80% of the freshwater, 10% of the available energy, and half of the land annually.

Figure 1.11: Every year, the EU wastes almost 88 million tons of food, with more than half of it originating from households. However, we can all help by altering the way we purchase, prepare, and plan our meals to prevent wasting food, money, and resources.

Source: https://www.eufic.org/en/food-safety/article/how-to-reduce-food-waste-at-home

Organic waste, predominantly food, is the second most common component of landfills, and landfills are the third most significant source of methane emissions. Because of its ability to absorb the sun's radiation and hence warm the atmosphere, methane is a key contributor to global warming. In the United States, 30 to 40% of food is wasted, amounting to up to 20 pounds of food per person every month. That translates to $165 billion in food waste every year in the United States (Qiao et al., 2018).

The food industry and consumers are the primary contributors to food waste in the United States. Waste happens at every stage of the food supply chain, including on the farm, as well as with packers, manufacturers, wholesalers, and retailers. Some of it is the product of financial factors, some of it is the consequence of managerial issues, and some of it is just the result of discarding things that aren't flawless in appearance.

Consumers may waste food due to a misinterpretation of what the terms on item date labels indicate, as well as confusion about how to store perishable items. Consumer food waste is reported to be 20 percent due to date labeling misunderstanding. The purpose and significance of date labels on packaged goods are sometimes misunderstood by customers. Consumer food waste is estimated to be 20 percent due to date labeling misunderstanding. Producers are not mandated by law or regulation to provide quality-based date labels on food packages, with the exception of baby formula. In the United States, there are no consistent or widely recognized definitions for open dating (calendar dates) on food labels. As a result, there are many different terms for product dating.

The FDA supports the food industry's attempts to make "Best if Used By" the standard phrase for indicating when a product's quality and freshness will be at their best. Consumers should inspect items that have passed their "Best if used by" date for signs of deterioration. Consumers should avoid ingesting items that have altered coloration, texture, or consistency significantly (Piggott & Marsh, 2004). If you have any issues or questions about the purity, quality, or labeling of the packaged goods you buy, you should contact the business that made them. Many packaged meals have contact information for the firm on the package.

Figure 1.12: Nearly a fifth of all food waste occurs during manufacturing and processing in affluent countries. Aside from the environmental effect of food processing waste, businesses have a strong financial motivation to eliminate food waste in their operations. Reduced industrial food waste may be a need in order to remain competitive in a changing market, given shrinking profit margins and increasing demand to increase production in order to feed an increasing world population.

Source: https://www.sesotec.com/apac/en/resources/blog/how-better-food-safety-technology-can-help-minimise-waste

Date labels are applied by manufacturers at their discretion and for a number of reasons. The most popular method is to tell customers and merchants the date by which the product will keep its desirable quality and flavor. The food industry is moving toward more consistent date labeling methods. However, for the time being, buyers may encounter other words for product date, such as Sell By, Best By, Expires on, and so on.

1.7 FOOD SAFETY AND SUSTAINABLE FOOD PRODUCTION

Because of the harmful consequences they have on the environment, several food production techniques have already attracted resistance from various populations. Food production techniques that not only safeguard the customer but also the environment is part of good food safety measures. Food safety regulation regulates practices such as reducing synthetic fertilizers that can seep into food items. Furthermore, food safety measures involve the provision of safe drinking water as well as the reduction of air, sewage, and other pollutants that have a significant impact on the environment. These activities are quite beneficial. Because water is a fundamental element in practically all processes, contaminated water may quickly damage all other substances and enter the human body.

Customers expect the food they buy to be safe. Unfortunately, as we have seen with several breakouts and large-scale recalls encompassing anything from leafy greens to meat to berries, this is not always the case. Millions of pounds of goods are withdrawn from the market, withdrawn from commerce, and destroyed in landfills as a result of these major food safety concerns. According to data from the Environmental Protection Agency, 75% of all food waste was disposed of in landfills, 18% was combusted with energy recovery, and just 6% was composted (Pressman et al., 2020).

Sustainability, like food safety, affects the whole supply chain, and everyone participating in the food manufacturing process may adopt sustainable practices. This includes everything from production agriculture to the use of organic and inorganic resources, as well as recycling of waste, packaging, composting, and a variety of other practices used across the supply chain during manufacture and distribution, as well as at the retail and consumer levels to meet future production requirements.

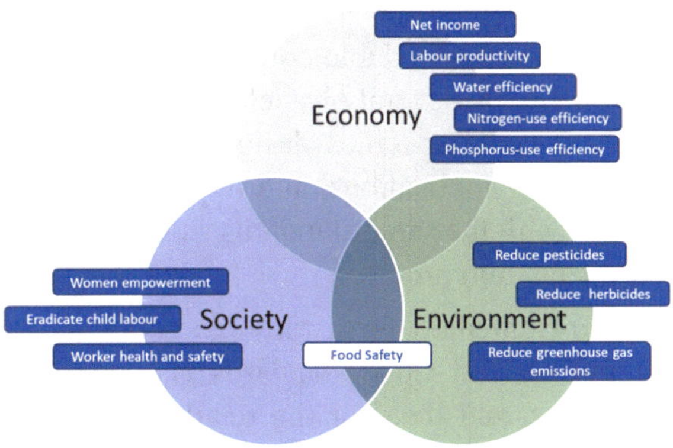

Figure 1.13: The key to ensuring food safety and long-term sustainability is to encourage developed-country governments to maintain tightening import rules. There is still space to improve rules that improve the food supply chain's safety and ensure more sustainable development. It remains to be seen if governments all across the world rise to the occasion and take action.

Source: https://globalfoodsafetyresource.com/sustainability-and-food-safety-whats-the-connection/

Despite the time and effort invested in sustainable practices across the supply chain, some challenges persist, and the aim of providing a safe product does not connect with sustainable practices. There are examples of how modifying one issue, such as food safety, might cause another possible concern. Sustainability is defined as development that meets current needs without jeopardizing future generations' capacity to satisfy stated or implied needs, according to a widely accepted definition of sustainability authored years ago in the landmark report "Our Common Future: Report of the World Commission on Environment and Development." By definition, sustainable practices are not sustainable if public health outcomes and food safety or are harmed as a result of their implementation.

Single-use plastics are under fire, and several authorities are considering outlawing their usage in specific industries. One that our business is familiar with is for food storage and servicing. Alternative materials, such as glass or biodegradable containers, can be used in some instances. To be able to fill the function of single-use plastics, more product development is required (Pollard et al., 2016). Many customers request that they be allowed to bring their own reusable containers to stores where they may buy things in bulk, such as salad or hot bars, or other prepared foods sections. This appears to be a straightforward solution, but there are other obstacles in the way, including issues with tare weights and other point-of-sale impediments.

Refilling customer-owned recyclable containers is addressed under the Food Code, however, there are several obstacles (FDA Food Code, Section 3-304.17(B)). The Food Code specifies design and construction criteria, restricts its usage to certain applications, and requires it to be washed and sterilized in accordance with the same standards as utensil cleaning in food outlets, as well as a verification process.

In summary, the only method for a reusable container to be utilized by consumers and fulfill Food Code standards is for the food business to authorize the container, and then for the retail establishment to clean, rinse, sterilize, and examine the container. In all realistic instances, these processes constitute a deterrent to utilizing consumer-provided reusable containers. This is a scenario in which the regulation, or framework for regulating, has not kept pace with consumer demand and advancements in sustainable practices. It's crucial that the container doesn't end up becoming a source of infection. Reusable containers, on the other hand, are routinely used in households and for noncommercial uses with no discernible public health consequences. If a compromise on safety can be struck, the usage of reusable containers appears to be a chance to promote sustainable practices and fulfill customer demand.

As of 2019, eight states have outlawed plastic bags, and several more jurisdictions have imposed limits or imposed charges on their use4. Bans on non-biodegradable bags, straws, and plastic are some of the variations. Bagging meat and poultry goods at grocery shops at the moment of purchase is one of the most efficient techniques of separating products. This has been shown in studies to lower contamination levels and safeguard consumers. Bans on plastic bags have an influence on food safety since they eliminate an efficient tool for isolating items from shop shelves and consumers' hands. Plastic bag exemptions for delis, seafood, poultry, meat, and produce departments have been negotiated with municipal and state legislators (Pappalardo et al., 2017).

Sustainability is concerned with environmental conservation, as well as human health and system economic sustainability. We must continue to work on our most difficult sustainability concerns, but we must also consider personal and social considerations, as well as the economic consequences. All elements must be taken into account, and there are several examples of unexpected effects that do more harm than good. We must challenge the food business to continue to develop and improve, but we must also safeguard it.

1.8 FOOD SAFETY AND GLOBALIZATION OF FOOD

For so long, globalization has been profoundly changing marketplaces and whole sectors that it is no longer a topic of discussion. The food and beverage sector, on the other hand, has been slower to embrace offshoring of production and sourcing options than other industries. Many individuals had the impression that their food was made in their own homes. Those days, too, are long gone.

Pharmaceutical companies operate in a highly regulated environment because governments understand that individuals are worried about the safety and efficacy of the substances they consume. Drug corporations are likewise aware that theirs is a trust-based industry. If the general public believes that a corporation cannot be trusted, and that it is ready to sell potentially harmful drugs, that company will not last long (Nesbitt et al., 2014). Producers of food and beverages are gradually accepting that they are part of the same trust economy. Unfortunately, many of them delayed much too long to embrace this reality—the PCA executives spring to mind quickly. They are now confronted with a slew of regulatory obligations and consumer expectations in terms of product safety and quality.

Figure 1.14: A diverse group of specialists is collaborating to enhance food safety, utilizing the most cutting-edge research and technology. Various government departments and agencies, including those responsible for public health, agriculture, education, and commerce, must coordinate and communicate with one another, as well as with civil society, particularly consumer organizations.

Source: https://www.one.org/international/blog/10-really-interesting-facts-about-food-safety/

Globalization opens up enormous potential for progress, but it also comes with it a slew of new issues that must be addressed. Tracing and tracking inputs and outputs across various borders and over a complicated supply chain has exposed food makers to far more risk and potential responsibility. The Global Food Safety Initiative (GFSI) is becoming a more widely acknowledged norm for food safety compliance, which has added to the administrative load on manufacturers.

According to CERT-ID, a renowned third-party food safety certification agency, an application must demonstrate a wide variety of competencies in order to be certified, including:

- Layout, working process, and segregation are all well-documented.
- Incident management mechanism in place for operations
- Certain handling criteria for specific goods, such as allergen-containing products and identity-preserved items, that are well-documented.

- Well-documented training methods, particularly employee training records.
- Senior management that is completely dedicated to putting the Standard's objectives into practice.
- A well-documented food safety strategy, as well as a Hazard Analysis and Critical Control Points (HACCP) analysis.
- Facilities with well-documented sanitation and Good Manufacturing Practices.

Many food and beverage manufacturers see these procedures as a bare minimum that must be fulfilled in order to meet GFSI and FSMA regulations. In fact, a dedication to food safety and quality may provide these businesses a competitive edge in the marketplace. However, the majority of these companies lack the organizational maturity to guide their employees through the task of altering how they handle food safety on the factory floor or in administrative operations (Meng & Doyle, 2002).

Many food and beverage manufacturers see these procedures as a bare minimum that must be fulfilled in order to meet GFSI and FSMA regulations. In fact, a dedication to food safety and quality may provide these businesses a competitive edge in the marketplace. However, the majority of these companies lack the organizational maturity to guide their employees through the task of altering how they handle food safety on the factory floor or in administrative operations. On a regular basis, new interpretations of the Food Safety Modernization Act's standards appear. With each passing year, the GFSI rules will continue to improve and mature. The new reality is constant change. It's pointless to ignore this. Embrace it and forge forth bravely into the future, or deny it and succumb to obsolescence and historical obscurity.

Global commerce in food and feed has been going on for a long time. Consumers in Europe have grown accustomed to a wide range of foods being accessible all year, regardless of seasonal or regional effects, thanks to imports from all parts of the globe (Beulens et al., 2005). Food safety is the product of a lot of efforts and thorough investigations all the way up the food chain, from the farmer to the consumer's plate. The standards applied to food safety and calculating residue limit levels are not as high everywhere as they are in the European Union, and other regions of the globe do not have as thorough labeling rules or as sensitive analysis techniques.

Our supply position is being shaped by "global sourcing." Foods and items that we consume no longer originate just from our region, neighboring nations, or specific international trading partners. Instead, there are enormous global networks for food and feed production and commerce. Food and feed company operators bear the major responsibility for food and feed safety. Within this vast network of economic contacts, identifying the producers is getting increasingly challenging. The traceability of foods, as well as the evaluation of their safety and quality, are severely hampered as a result.

Smaller producers in poor and emerging economies may find it difficult to comply with national food regulations in export markets. Many developing countries import a considerable portion of the food they consume. Some countries, such as the Pacific Islands, rely nearly totally on imports to meet their food needs. According to the FAO, occasional border inspections are no longer appropriate due to increased volumes of imported foods, variety of origin, and rising complexity of the technology utilized for traditional monitoring procedures.

Insurance may assist a corporation to transfer some of the remaining risk, in addition to a strong risk management approach to better assess and mitigate food safety concerns. Product recalls in the food business, as well as the potential for reputational harm, may have a devastating effect on a company's finance sheet (Carvalho, 2006). Worse, individuals may be harmed as a result of ingesting tainted food and seek compensation for their injuries. Some firms have been forced out of the market due to large-scale product recall actions and product liability claims, as well as a loss of trust and reputation.

Food safety plays a great role within the entire food chain more than we think it does. Food businesses are required to implement food safety management systems to protect the health of consumers from any foodborne disease. Beyond this concept, food safety contributes as well to important necessities in society as food security. When properly implemented, food safety practices can help detect and control food safety risks, contribute to economic prosperity, food security, and continuous and sustainable development. By providing safe food for consumption a continuous supply of food for everyone can be ensured. The many benefits of food safety in various aspects of life prove why food safety is important for everyone.

CHAPTER 2

THE HISTORY OF FOOD SAFETY

CONTENTS

2.1 INTRODUCTION

Food safety is about not getting sick from the food you eat. It is possible to get sick from food because it can be contaminated with germs like bacteria, viruses, or parasites. The bacteria that cause food-borne illness are everywhere and they are especially common in raw meat, poultry, seafood, and eggs (Mensah & Julien, 2011). Although these bacteria are all around us, they can't make you sick unless they get into your body. If you don't handle or prepare your food carefully and safely, germs can spread from one food to another and make you sick. This is called cross-contamination.

Figure 2.1: An image showing the four main steps to food safety.

Source: https://www.google.com/imgres?imgurl=https%3A%2F%2Fwww.cdc.gov%2Ffoodsafety%2Fimages%2Fflexslider%2F4-steps-banner re-vised.jpg%3F_%3D45978&imgrefurl=https%3A%2F%2Fwww.cdc.gov%2Ffoodsafety%2Fkeep-food-safe.html&tbnid=vhqVdyyg353ZUM&vet=12ahUKEwje7LTAr-b2AhVblKQKHeWYAG8QMygAegUIARDXAQ..i&docid=YKjKNQ5_oYkEaM&w=890&h=387&q=food%20safety&ved=2ahUKEwje7LTAr-b2AhVblKQKHeWYAG8QMygAegUIARDX-AQ

In the 1800s, the term safety was rarely used, and when it was, it was often used in a way that we can't even imagine today. Back then, food safety was not necessarily even something that people thought about. Instead of food safety, they looked at food cleanliness, which is kind of what we would think of as how hygienic and sanitary the food is. In other words, if the food was clean, it was safe to eat. But this idea changed with the 1906 publication of Upton Sinclair's novel The Jungle. It exposed unsanitary conditions in Chicago meat-packing factories and brought issues of hygiene and safety to light. People were shocked, and concerned, that their meat wasn't being handled in sterile conditions (McLaughlin et al., 1999). So laws were passed to keep meatpacking plants cleaner and safer for consumers. And so the term food safety came about, and it stuck around. This shift in definition also led to a change in focus on what people needed to do to keep their food safe: instead of looking at how consistent they are with washing their hands, they should be consistent with the cleanliness of the food they consume.

2.2 GENERAL HISTORY OF FOOD SAFETY

Food is one of the most essential gifts that we receive from nature. It provides us with energy and allows us to grow into healthy adults. A well-balanced diet is crucial for our survival, as well as for the survival of our children, who need the right amount of nutrients in order to develop and grow properly. But what happens when food is contaminated or unsafe to eat. Well, this can lead to serious health issues like food poisoning which has been affecting humans since ancient times. The first recorded case of a foodborne illness was found in 1799 BC. A physician described how people suffered from stomach aches after consuming spoiled meat. He also noted that they had symptoms such as nausea, vomiting and diarrhea due to bacterial growth inside their bodies due to eating rotten food.

Figure 2.2: An image showing the main threats to food safety.

Source: https://www.google.com/imgres?imgurl=https%3A%2F%2Fwww.foodsafetynews.com%2Ffiles%2F2021%2F02%2Fdreamstime_food-safety_foodborne.jpg&imgrefurl=https%3A%2F%2Fwww.foodsafetynews.com%2F2021%2F05%2Fwho-opens-consultation-on-draft-food-safety-strategy%2F&tbnid=yZOxljJCJ9psnM&vet=12ahUKEwje7LTAr-b2AhVblKQKHeWYAG8QMygDegUIARDdAQ..i&docid=ylLNcSzh2j_w_M&w=1197&h=699&q=food%20safety&ved=2ahUKEwje7LTAr-b2AhVblKQKHeWYAG8QMygDegUIARDdAQ

The history of food safety begins in ancient times when rulers made laws about what foods were safe to sell in markets. In early societies like Rome, the city of Rome itself was largely responsible for establishing standards for food quality and safety. Later on, as trade routes expanded and began to connect societies across the world with each other, new foods started to be introduced to different cultures. As a result, new health issues arose, and people began taking more interest in understanding how their food was grown and produced. In 1856, the London Water Supply Company began testing water samples for bacteria content. This was an important development because it represented an acknowledgment that there was some measurable level of contamination that could be identified in food or drink products. This discovery led people to begin analyzing other substances like milk and meat to see if they could find any levels of impurities or contaminants within them as well. The Industrial Revolution brought on numerous changes in how people prepared their meals and stored their foods for consumption (Chandrapala et al., 2012).

In the 1900s, there were a couple of big advances in making our food supply safer: one was called canning, and involved putting food in air-tight containers called cans. It helped companies keep food fresh for weeks or months at a time. The other advance was freezing food. In 1960, the Food and Drug Administration (FDA) introduces two new food additives: sodium nitrite and sulfur dioxide, used to preserve foods and prevent botulism poisoning.

Nowadays, you may take food safety for granted, but that wasn't always the case. Back in times of ancient Egypt, around 3100 BC, they had no idea what we know now about pasteurization, irradiation, and other ways of preventing the spread of bacteria. In fact, most early civilizations had no understanding of bacteria at all, let alone foodborne illnesses. But when you consider that around one in six people in the world get sick from consuming contaminated food or water every year, it is not hard to see the importance of food safety. These days, we're lucky enough to live in a time where we can minimize our risk of getting sick by following simple rules like don't eat raw dough, and wash your hands before preparing food.

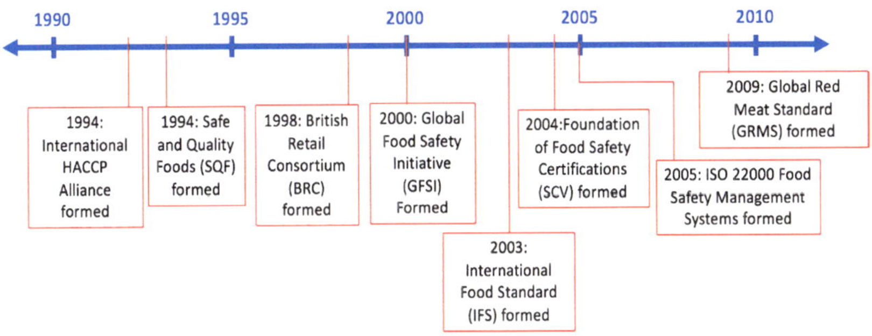

Figure 2.3: An image showing a summary of the general history of food safety.

Source:https://www.google.com/imgres?imgurl=https%3A%2F%2Fwww.researchgate.net%2Fpublication%2F327327154%2Ffigure%2Ffig2%2FAS%3A665515550859275%401535682680686%2FTimeline-of-introduction-of-food-safety-management-systems.png&imgrefurl=https%3A%2F%2Fwww.researchgate.net%2Ffigure%2FTimeline-of-introduction-of-food-safety-management-systems_fig2_327327154&tbnid=59qYFV3-v9ic6M&vet=12ahUKEwiBlfyHsOb2AhVWtioKHac0CKgQMygCegUIARC7AQ..i&docid=EccS7_yy8WsP8M&w=850&h=284&q=history%20of%20food%20safety&ved=2ahUKEwiBlfyHsOb2AhVWtioKHac0CKgQMygCegUIARC7AQ

Since the mid-1970s, specialists have ceaselessly sounded such cautions approximately holes within the government nourishment security framework. In 1998, the National Foundation of Sciences cited at times bygone nourishment statutes that obstruct science-based decision-making. In 1999, the Government Responsibility Office (GAO) cautioned that our fragmented framework was not created beneath any sound arrangement but was fixed together over a long time to address particular wellbeing dangers from specific nourishment products. Thirty years back, today's most fearsome threats–such as *Campylobacter pylori*, *Listeria monocytogenes* and *E. coli* were ignored by most individuals even the ones in the medical sector or yet to be discovered (Chen et al., 2020). Indeed nowadays, endless puzzling microbes are hiding. It has only been recently that we have become more aware of the importance of food safety. People have become more aware of what goes into their food and how it is prepared. This can help to keep you from getting sick from your food or from getting sick from dirty dishes in restaurants.

2.3 DATED HISTORY OF FOOD SAFETY

It seems like food safety has always been a concern. Ancient Babylonians worried about it, and so did the Romans. In fact, in ancient civilizations, religious leaders were responsible for ensuring food safety. They were the ones who were supposed to know how to keep food from spoiling, as well as how to diagnose and treat food poisoning. The first municipal health inspector was actually appointed in 1805, almost 2,000 years after the first mention of food safety. The position was created by Maximilien Robespierre, and the inspector's job was to ensure that the meat being sold at markets wasn't rotten. Throughout the 19th century, many more public health positions were created all over Europe and America. Food law became increasingly strict throughout this period of time (Cody et al., 2012).Today, we have many more regulations than ever before, both governmental and private, to ensure that our food is safe to eat. We can be confident that our food is safe from contamination, as well as free from additives like preservatives that might cause us harm in the long term.

2.4 THE USDA AND FDA FOOD SAFETY REGULATIONS

In 1862, the United States Department of Agriculture and the Food and Drug Administration was established. They were both parts of the Morrill Act, which was signed by President Abraham Lincoln. These acts established regulations to make sure that food is safe and works to improve it. Since then, they have continued to work on making food safer and better in America. It wasn't until 1906 that the Pure Food and Drug Act put the FDA in charge of monitoring food safety in the US. It regulates everything from food packaging to labeling, including nutrition facts, and it also has jurisdiction over additives like flavorings, colorings, preservatives, and processing agents. The USDA's main focus is on meat, poultry, and eggs. The FDA also oversees dietary supplements and other items like cosmetics and over-the-counter medications. Over time, those agencies have grown and reached across several areas of food safety regulation. The FDA is responsible for inspecting restaurants' kitchens for cleanliness and checking for illegal chemical additives or incorrect labeling information on packaged foods. The USDA regulates meat processing plants and inspects all meat sold in grocery stores to ensure it is safe to eat.

Figure 2.4: An image showing the differences between USDA and FDA.

Source: http://foodsafety.merieuxnutrisciences.com/2018/06/28/understanding-key-usda-fda-food-labeling-differences-part-one/

The FDA has no such official duties. In fact, their only job is to make sure products don't make false claims about their health benefits or risk of adverse events after consumption-which means that they can only do so much when it comes to enforcing food safety standards and labeling requirements. While this may seem like an inadequate amount of oversight at first glance, consider how little information consumers are given about what goes into their meals every day and how often those ingredients aren't even listed on labels (MacLaurin, 2001). This lack of transparency clearly needs reform because it allows companies like McDonald's and Nestle to manipulate people into thinking they're eating healthy foods when really they're just getting fatter due to a lack of regulation around labeling requirements and food safety standards. The FDA could easily fix this problem by implementing stricter rules around both topics so everyone would know exactly what they're consuming before putting anything into their mouths.

2.5 PURE FOOD AND DRUG AND FEDERAL MEAL INSPECTION ACT

Food safety is a top priority for businesses in the food industry. To make sure that consumers are able to trust their products, food manufacturers have to follow strict federal guidelines. The Pure Food and Drug Act of 1906 and the Federal Meat Inspection Act of 1906 were two groundbreaking pieces of legislation that paved the way for the modern food safety guidelines that we use today. The Pure Food and Drug Act of 1906 and the Federal

Meal Inspection Act of 1907 were both sponsored by Senator Albert J. Beveridge, who was a Republican from Indiana. These laws were passed by Congress in response to a period of industrialization, urbanization, and immigration. These two pieces of legislation prohibited the manufacture, sale, or transportation of adulterated or misbranded foods and drugs across state lines.

Figure 2.5: An illustration of the 1906 act being applied to meat inspection.

Source: https://www.google.com/imgres?imgurl=https%3A%2F%2Fimage1.slideserve.com%2F2584752%2Fthe-pure-food-and-drug-act-meat-inspection-act-1906-n.jpg&imgrefurl=https%3A%2F%2Fwww.slideserve.com%2Flester%2Fthe-pure-food-and-drug-act-meat-inspection-act-1906&tbnid=GYAXwHAWgrX94M&vet=12ahUKEwjmnIX7sOb2AhVmXvEDHaW6BT4QMygFegUIARC6AQ..I&docid=IXv2P8CunbSLXM&w=720&h=540&q=Pure%20Food%20and%20Drug%20and%20Federal%20Meat%20Inspection%20Act&ved=2ahUKEwjmnIX7sOb2AhVmXvEDHaW6BT4QMygFegUIARC6AQ

The Federal Meat Inspection Act was passed a year later, in response to Upton Sinclair's novel "The Jungle." Sinclair's novel highlighted the unsanitary conditions in meatpacking plants. As a result of the novel, Congress passed the Federal Meat Inspection Act, which required that all meat products be inspected and labeled before being sold.

The new laws were enforced by the Bureau of Chemistry in the Department of Agriculture, which was later renamed the Food and Drug Administration. At the time, food manufacturers did not have to list ingredients on their products, so consumers had no way of knowing what they were purchasing. These new laws changed that; now manufacturers had to list ingredients on their products in order for them to be sold interstate. The FDA also started inspecting food shipments before they crossed state borders. The Pure Food and Drug Act helped establish America's food safety standards today by giving the FDA regulatory power over food companies' manufacturing processes as well as interstate shipping practices (Diasolua et al., 2002).

In 1938, the Federal Food, Drug and Cosmetic Act replaced the Pure Food and Drug Act. Among other things, the 1938 law changed labeling requirements to include nutrition information on labels, established new standards for drugs such as antibiotics, and required testing of cosmetics before they were sold to consumers.

2.6 PURE FOOD AND DRUG REVISED ARTICLE IN 1938

The Pure Food and Drug Act of 1906 was the first of a series of significant consumer protection laws which was enacted by Congress in the 20th century. It was likely to have been instigated by muckraking journalists, like Upton Sinclair and Samuel Hopkins Adams. Their reports on the unsafe production of food and drugs prompted President Theodore Roosevelt to demand a Congressional investigation. By the 1920s, almost all states had adopted pure food laws of their own which mirrored the federal law. This meant that individual state governments were now able to take legal action against companies that sold contaminated foods or drugs.

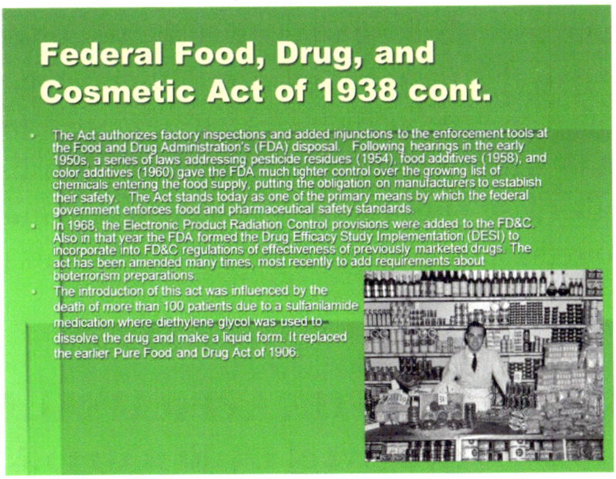

Figure 2.6: An image showing the continuation of the 1938 act.

Source: https://www.google.com/imgres?imgurl=https%3A%2F%2Fimages.slideplayer.com%2F24%2F7262099%2Fslides%2Fslide_3.jpg&imgrefurl=htt

ps%3A%2F%2Fslideplayer.com%2Fslide%2F7262099%2F&tbnid=h4eM6ZxGpIUyrM&vet=12ahUKEwi9qK-wseb2AhVpi_0HHSUID4gQMygBegUIARCUAQ..I&docid=HcFpB4lb2k0m0M&w=960&h=720&q=Pure%20Food%20and%20Drug%20Revised%20Article%20in%201938&ved=2ahUKEwi9qK-wseb2AhVpi_0HHSUID4gQMygBegUIARCUAQ

Unfortunately, there were many loopholes in the Pure Food and Drug Act which allowed companies to mislabel products without being punished. For instance, a company could still label its product as containing alcohol even if it only contained traces of it. Or a company could claim an additive was "harmless" even if it wasn't true. In 1938, President Franklin Delano Roosevelt introduced new legislation called The Food, Drug, and Cosmetic Act (FDCA). The FDCA made amendments to the 1906 Pure Food and Drug Act which closed some loopholes and expanded the definition of "adulterated" and "misbranded" foods.

The revised act had several requirements which had to be met: Food shall be clean and free from filth and shall not be adulterated or misbranded; It shall be filed with the Commissioner of Food and Drugs a true copy of the label he proposes to use on such article; No food shall have any added poisonous or deleterious substance which may render such article injurious to health, but the term "deleterious substance" shall not include common salt, spices, flavorings or chemicals used in the preparation of food in accordance with any practice which has been or may hereafter be established by regulations of the Secretary; but no substance which is intentionally included in food for preservative purposes shall be deemed to be poisonous or deleterious within the meaning of this paragraph if it is within the quantity prescribed by regulations issued by the Secretary as safe for such purposes; and, no food shall consist in whole or in part of any filthy, putrid, or decomposed substance or be otherwise unfit for food; nor shall any substance be mixed therewith, so as to reduce or lower or injuriously affect its quality, strength, or purity (Mou & Lin, 2014).

2.7 PROCEDURES FOR THE APPRAISAL OF THE TOXICITY OF CHEMICALS IN FOOD OF 1949

The Procedures for the Appraisal of the Toxicity of Chemicals in Food of 1949 were developed by the World Health Organization (WHO) to determine whether or not certain chemicals are safe to consume, and at what levels. All edible goods sold in the United States are subject to the provisions of the

Procedures for the Appraisal of the Toxicity of Chemicals in Food, a piece of legislation that requires all foods containing chemicals to be tested for toxicity.

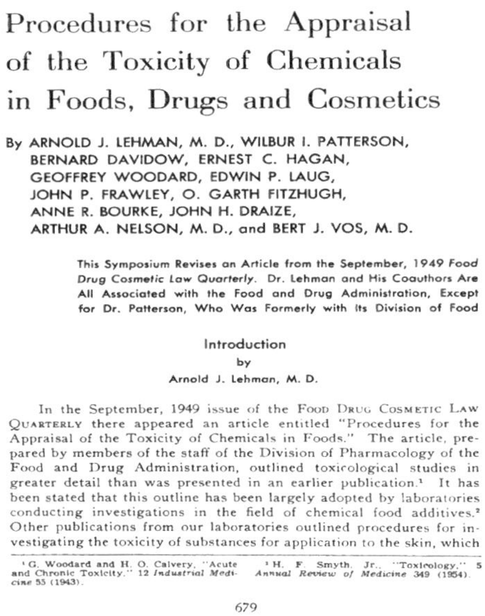

Procedures for the Appraisal of the Toxicity of Chemicals in Foods, Drugs and Cosmetics

By ARNOLD J. LEHMAN, M. D., WILBUR I. PATTERSON, BERNARD DAVIDOW, ERNEST C. HAGAN, GEOFFREY WOODARD, EDWIN P. LAUG, JOHN P. FRAWLEY, O. GARTH FITZHUGH, ANNE R. BOURKE, JOHN H. DRAIZE, ARTHUR A. NELSON, M. D., and BERT J. VOS, M. D.

This Symposium Revises an Article from the September, 1949 *Food Drug Cosmetic Law Quarterly*. Dr. Lehman and His Coauthors Are All Associated with the Food and Drug Administration, Except for Dr. Patterson, Who Was Formerly with Its Division of Food

Introduction
by
Arnold J. Lehman, M. D.

In the September, 1949 issue of the FOOD DRUG COSMETIC LAW QUARTERLY there appeared an article entitled "Procedures for the Appraisal of the Toxicity of Chemicals in Foods." The article, prepared by members of the staff of the Division of Pharmacology of the Food and Drug Administration, outlined toxicological studies in greater detail than was presented in an earlier publication.[1] It has been stated that this outline has been largely adopted by laboratories conducting investigations in the field of chemical food additives.[2] Other publications from our laboratories outlined procedures for investigating the toxicity of substances for application to the skin, which

[1] G. Woodard and H. O. Calvery, "Acute and Chronic Toxicity," 12 *Industrial Medicine* 55 (1943).

[2] H. F. Smyth, Jr., "Toxicology," 5 *Annual Review of Medicine* 349 (1954).

679

Figure 2.7: An image of the Appraisal of the Toxicity of chemicals in food act.

Source: https://www.google.com/imgres?imgurl=x-raw-image%3A%2F%2F%2F7488acff291e6fba00f780954d397d33f29fe4d-4f52667a0c8bdb831cd44e3d2&imgrefurl=https%3A%2F%2Fwww.jstor.org%2Fstable%2F26655418&tbnid=um8eKf-siHTSnM&vet=12ahUKEwi5qp7aseb2AhWdYfEDHeGmDCQQMygBegQIARAe..i&docid=D7oUzFCGQ-UJFM&w=1600&h=2462&q=Procedures%20for%20the%20Appraisal%20of%20the%20Toxicity%20of%20Chemicals%20in%20Food%20of%201949&ved=2ahUKEwi5qp7aseb2AhWdYfEDHeGmDCQQMygBegQIARAe

2.8 FEDERAL FOOD, DRUG, AND COSMETICS ACT OF 1938 AMENDED IN 1958

The Federal Food, Drug, and Cosmetic Act (FFDCA) was signed into law by President Franklin Delano Roosevelt on June 25, 1938. It was the most comprehensive federal food and drug law in U.S. history, giving the Food and Drug Administration (FDA) authority over foods, drugs, and cosmetics. Under the Act, the FDA was required to establish uniform safety standards for human and animal drugs. In 1958, the FFDCA was amended to include provisions requiring efficacy and safety determination of new animal drugs before they could be marketed. The law has been amended many times since 1938, but its underlying principles remain unchanged: In order to prevent harm to consumers, FDA must ensure that drugs are safe as well as effective; that food is safe as well as wholesome; that color additives are safe; those cosmetics are safe; and that medical devices are safe as well as effective.

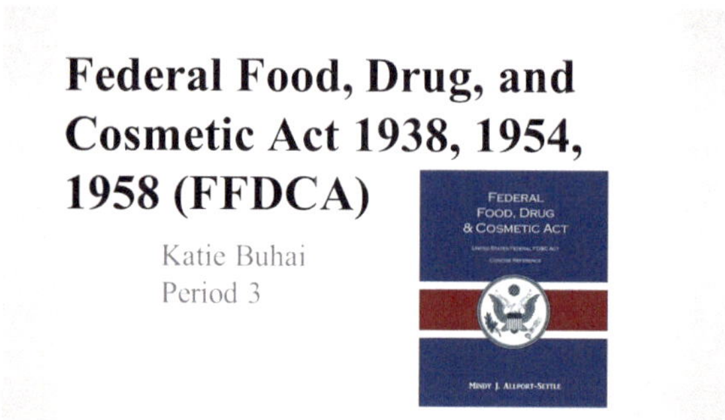

Figure 2.8: An image showing the Amended act of 1938 in 1958.

Source: https://www.google.com/imgres?imgurl=https%3A%2F%2Fslideplayer.com%2F7262317%2F24%2Fimages%2Fslide_1.jpg&imgrefurl=https%3A%2F%2Fslideplayer.com%2Fslide%2F7262317%2F&tbnid=1KkQ6jb7gBNzqM&vet=12ahUKEwiA-Y-Jsub2AhVOEsAKHZHpCncQMygAegUIARCQAQ..i&docid=JQigshFzRNTRHM&w=960&h=540&q=Federal%20Food%2C%20Drug%2C%20and%20Cosmetics%20Act%20of%201938%20Amended%20in%201958&ved=2ahUKEwiA-Y-Jsub2AhVOE-sAKHZHpCncQMygAegUIARCQAQ

A guide to a nutritional label based on this law includes the following: Serving Size: which tells you how much there is in one serving. For example, if the package says there are two servings per container, and you eat the whole thing, then you've eaten twice as many calories as the label states. Servings per Container, which tells you how many servings there is total in the package. Calories, which tell you how much energy is in each serving. Nutrients, whereby the nutrients listed here have been determined to be most relevant to your daily diet. They include fat, cholesterol, sodium, carbohydrates, fiber, sugars, protein and vitamins or minerals. Percent Daily Value, which shows you how much of each nutrient is in one serving compared to your daily needs. If one serving of macaroni and cheese (EFSA et al., 2020).

2.9 CONSUMER BILL OF RIGHTS OF 1962

The Consumer Bill of Rights of 1962 is a food safety regulation that emphasizes the importance of transparency in the manufacturing and distribution of food products. One important aspect of this regulation that has helped to ensure food safety is the requirement for manufacturers to provide clear and easy-to-understand label information on all food items. Not only does this make consumers aware of potential allergens or other caveats, but it also helps them to understand whether the products they are buying are healthy options. It also allows them to make more informed decisions about the foods they purchase.

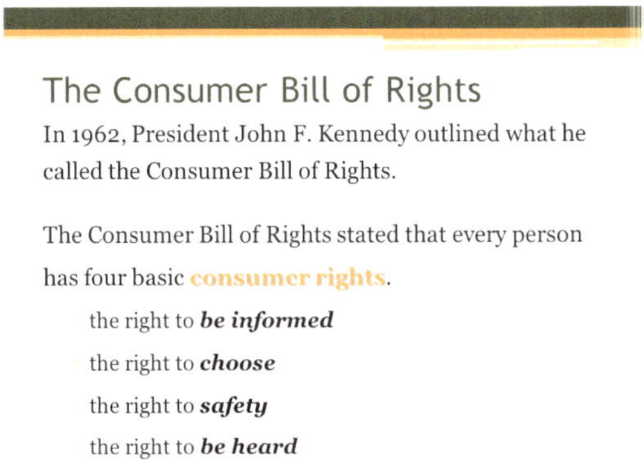

Figure 2.9: An image showing a summary of the consumer bill of rights.

Source: https://www.google.com/imgres?imgurl=https%3A%2F%2Fimages.slideplayer.com%2F13%2F3934084%2Fslides%2Fslide_5.jpg&imgrefurl=https%3A%2F%2Fslideplayer.com%2Fslide%2F3934084%2F&tbnid=V5ZzVaZ6TEfP_M&vet=12ahUKEwi2xL-0sub2AhWJ-qQKHWrhC0oQMygAegUIARC2AQ..I&docid=cSWZJrTSxLmZaM&w=960&h=720&q=Consumer%20Bill%20of%20Rights%20of%201962&ved=2ahUKEwi2xL-0sub2AhWJ-qQKHWrhC0oQMygAegUIARC2AQ

The regulation also requires that manufacturers document their processes and methods, which ensures that their products meet industry standards for safety and quality. This gives consumers peace of mind that the foods they buy are safe to eat. This bill is an important step in ensuring that our food stays safe from pathogens like *E. coli* or Salmonella. It also protects against genetically modified organisms (GMOs), which are often found in foods like corn or soy beans.

The four basic rights it established are the rights to safety, to choose, to be informed, and to be heard.

2.10 FAIR PACKAGING AND LABELING ACT ENACTED OF 1967

In 1966, Congress passed the Fair Packaging and Labeling Act, which required manufacturers to put labels on their products that are accurate, informative, and not misleading. They also mandated that all consumer products have standardized containers so that customers can compare prices by volume or weight. In 1967, the Fair Packaging and Labeling Act were passed to protect consumers from false or misleading labeling. The act addresses what needs to be on a label, where it should be placed, and how large it should be in relation to the product.

Figure 2.10: An image showing the Logo for the Fair Packaging and Labeling Act Enacted of 1967.

Source: https://www.google.com/imgres?imgurl=https%3A%2F%2Fwww.abbreviations.com%2Fimages%2F1420842_FPLA.png&imgrefurl=https%3A%2F%2Fwww.abbreviations.com%2Fterm%2F1420842&tbnid=ccVodz-hPZoaiM&vet=12ahUKEwjc5Z7Wsub2AhUrEmMBHSbTAe4QMygBegQIARBB..i&docid=uedwAmJhSsLgRM&w=500&h=300&q=Fair%20Packaging%20and%20Labeling%20Act%20Enacted%20of%201967&ved=2ahUKEwjc5Z7Wsub2AhUrEmMBHSbTAe4QMygBegQIARBB

Since the passage of the Fair Packaging and Labeling Act in 1967, consumers have had the right to know what they are buying. The act requires certain types of product information be displayed on packaging, including the identity of the product; the product's volume or weight; net quantity or count; the manufacturer's name and location; and ingredients listed in descending order by ingoing weight.

2.11 CENTERS FOR DISEASE CONTROL OF 1970

The Centers for Disease Control (CDC) began keeping records on cases of foodborne illness in 1970. Foodborne illness is a growing problem. In fact, since the Centers for Disease Control (CDC) began keeping records in 1970, the number of foodborne illnesses has steadily increased. Today, nearly half of all Americans will suffer from a foodborne illness at some point in their lives, and over 120,000 will require hospitalization. Every year, one out of every six Americans gets sick from contaminated food or beverages (Lytton, 2017). There are many reasons why incidents of foodborne illness are on the rise, including: increased U.S. and global trade of food products that may contain pathogens; New technologies that can accidentally introduce contamination into the food production process; the evolution of known pathogens, resulting in new strains that are more virulent and resistant to antibiotics than ever before. With so much at stake, it is important to monitor your products closely to ensure they are free from contamination.

Figure 2.11: An image showing the center for disease control.

Source: https://www.google.com/imgres?imgurl=https%3A%2F%2Fi.pinimg.com%2Foriginals%2F30%2Fd6%2F42%2F30d642f924e8623730c8069c4a975189.jpg&imgrefurl=https%3A%2F%2Fwww.pinterest.com%2Fpin%2F290763719662171365%2F&tbnid=17fy4VUfDkF_CM&vet=12ahUKEwjl--WPs-b2AhXylosKHYjvBoIQMygAegUIARC6AQ..i&docid=QUnHphk8n5UBRM&w=960&h=660&itg=1&q=Centers%20for%20Disease%20Control%20of%201970&ved=2ahUKEwjl--WPs-b2AhXylosKHYjvBoIQMygAegUIAR-C6AQ

Foodborne illness can be caused by bacteria, viruses, and parasites. The CDC estimates each year, 48 million people will get sick from foodborne illnesses. That's one in six Americans who contract a foodborne illness each year. One of the main problems is that people don't know that they have a foodborne illness. The symptoms are often mild and people may assume they have something else, like the flu. The US Department of Agriculture (USDA) and the FDA work hard to keep our food safe. From food production to food preparation to serving foods, many government organizations are keeping watch over the process of getting our food from the farm to our plates (Li et al., 2017). Unfortunately, there are still some instances where unsafe foods make it onto our plates. When this happens, it's important to report it so the problem can be fixed and you can get help if you need it.

2.12 FIRST MAJOR FOOD RECALL IN THE U.S OF 1973

In 1973, the United States Congress passed the first major food safety law in1938. The law established a consumer-protection agency within the Food and Drug Administration that would be responsible for inspecting all domestic and imported foods, except meat and poultry. It also required that food processors adhere to federal standards of sanitation, good manufacturing practices, and hazard analysis and critical control point (HACCP) systems.

Figure 2.12: An image showing the trade and food standards.

Source:https://www.google.com/imgres?imgurl=x-raw-image%3A%2F%2F%2F1ad258875529a6a47c8761eb22044a3fd8f07ebf08eb-72fa60f14eeb1dbecd44&imgrefurl=http%3A%2F%2Fwww.fao.org%2F3%2Fi7407e%2Fi7407e.pdf&tbnid=pKYpVYdY0hrZsM&vet=12ahUKEwirmozZs-b2AhWHuyoKHTI2DQEQMygLegQIARA0..i&docid=CMke9l2_4xypnM&w=1600&h=2263&q=First%20Major%20Food%20Re-call%20in%20U.S%20of%201973&ved=2ahUKEwirmozZs-b2AhWHuy-oKHTI2DQEQMygLegQIARA0

In 1973, the first major food recall was announced in the United States. The culprit was Aflatoxin which is a toxin that occurs when foods are stored at high temperatures and humidity. In this case, it was found in peanut butter products. This event caused the Food and Drug Administration to create the Food Safety and Inspection Service to ensure that all food products are safe for human consumption.

The FDA is in charge of enforcing food safety laws. The FDA has several programs that are designed to keep our food supply safe, including the regulation of produce and seafood, as well as meat and poultry. When it comes to keeping food safe for consumers to eat, the most important thing for both producers and consumers is education about proper handling practices and hygiene standards. All three of these things play a role in preventing contamination from occurring in the first place.

2.13 FOOD SAFETY AND QUALITY SERVICE OF 1977

In 1977, the United States passed the Food Safety and Quality Service Act (FSQA), which is an extensive food safety law that regulates the production of meat, poultry, and eggs. It set a number of regulations, including making sure all restaurant staff is trained in food safety, using safe and clean ingredients, and more. This means that farms and factories have to adhere to strict regulations about how they treat their animals and how they process their products. For example, raising cows for milk or beef, one can't use hormones or other growth agents to make your milk or beef more profitable. One also has to provide a clean environment for your animals and make sure they are fed a nutritious diet. If one is processing meat into sausage or ground beef, you can't put anything in there that isn't meat, you can't mix in your leftover milk or yogurt.

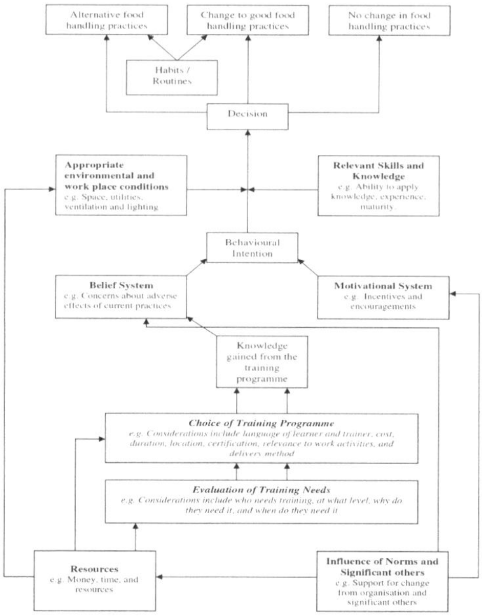

Figure 2.13: An image showing a food hygiene training manual.

*Source: https://www.google.com/imgres?imgurl=https%3A%2F%2Fwww.researchgate.net%2Fprofile%2FPhillip-Seaman%2Fpublication%2F223695316%2Ffigure%2Ffig1%2FAS%3A340639322525714%401458226151595%2FProposed-considerations- for-effective-food-hygiene-training_Q320.jpg&imgrefurl=https %3A%2F%2Fwww.researchgate. net%2Fpublication%2F223695316_The_ management_of_food_safety-the_role_of_food_hygiene_training_in_the_UK_service_sector&tbnid =Dvsj8MN0UAW_eM&vet =12ahUK Ewicut36s- b2AhUsif0HHZCzCLgQMygBegQIARA_..i&docid =FhxTADl8PeFAgM&w=320&h=320&q=Food%20Safety%20and%20Quality%20Service%20of%201977&ved=2ahUKEwicut36s-b2AhUsif0H-HZCzCLgQMygBegQIARA_*

In addition to food safety laws, we also have quality standards, standards that are voluntary but which many people choose to follow. These guidelines help determine how delicious our food will be by regulating everything from the amount of fat in our meat to the amount of sugar in our juice.

This is a food safety law that was passed as part of the United States Food and Drug Administration's efforts to ensure that the food available in grocery stores is safe for consumers. This law has been updated since it was passed, most recently in 2015, but the core tenets remain the same: ensuring the safety of food and other products available for human consumption.

2.14 FOOD SAFETY AND INSPECTION SERVICE OF 1979

The Food Safety and Inspection Service (FSIS) is the public health agency in the U.S. Department of Agriculture responsible for ensuring that the nation's commercial supply of meat, poultry, and egg products is safe, wholesome, and correctly labeled and packaged. The FSIS was established on March 4, 1979, as a result of a major food safety law passed by Congress in 1973 (Lam et al., 2013). The FSIS oversees the safety of meat, poultry, and processed eggs. In addition to federal inspection services, FSIS provides technical assistance to state meat and poultry inspection programs.

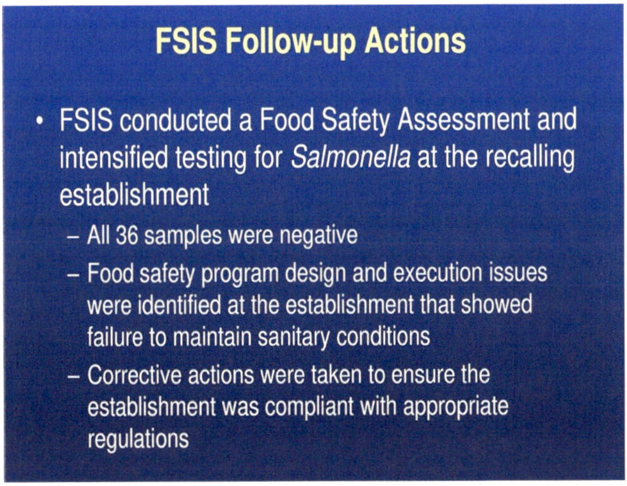

Figure 2.14: An image showing the FSIS follow-up activities.

Source: https://www.google.com/imgres?imgurl=https%3A%2F%2Fslideplayer.com%2Fslide%2F13279264%2F79%2Fimages%2F14%2FFSIS%2BFollow-up%2BActions.jpg&imgrefurl=https%3A%2F%2Fslideplayer.com%2Fslide%2F13279264%2F&tbnid=GtPJ-vEgIw8ABM&vet=12ahUKEwjbkJ6btOb2AhWJnhQKHczgBSwQMygBegQIARA8..i&docid=11hQReCAn675MM&w=1024&h=768&q=Food%20Safety%20and%20Inspection%20Service%20of%201979&ved=2ahUKEwjbkJ6btOb2AhWJnhQKHczgBSwQMygBegQIARA8

FSIS offers many services to consumers including information about food product recalls, inspection requirements for meat, poultry, and egg products, how to prevent foodborne illness with safe food handling tips, how to report a problem with a meat or poultry product, how to find an FSIS office near you and more.

2.15 PATHOGEN REDUCTION OR HACCP SYSTEMS LANDMARK RULE OF 1996

In 1996, the Food Safety and Inspection Service (FSIS) implemented a landmark rule designed to ensure the safety of meat and poultry products. The Pathogen Reduction or "HACCP" Systems rule marked the first time in history that meat processing plants were required to institute a food safety plan that was monitored by the federal government.

To ensure food safety, we also utilize HACCP systems. This stands for Hazard Analysis Critical Control Points, and it is the industry standard for ensuring food safety for our customers. The HACCP system itself was developed in response to an earlier crisis in food safety: In 1993, *E. coli* O157:H7 infection from undercooked Jack-in-the-Box hamburgers killed four children, sickened hundreds of people, and was linked to 500 million dollars in damages to the fast-food chain (King et al., 2017). The HACCP system mandated by the 1996 Act was designed to prevent similar situations in the future. Before HACCP, meat and poultry safety inspections were primarily reactive—they waited until there was a problem before they attempted to correct it. Even after HACCP systems were implemented, this approach remained an option for less technically advanced plants, but most major producers now use HACCP as their main method of ensuring food safety.

Under this model, every step of production is carefully recorded and monitored by both plant personnel and FSIS inspectors. The HACCP system was originally developed by the National Advisory Committee on Microbiological Criteria for Foods for NASA in the 1960s for use in their space programs. The rule was modified in 2012 to require salmonella testing for chicken parts and ground chicken.

2.16 FOOD AND DRUG MODERNIZATION ACT AMENDED IN 1997

The Food and Drug Modernization Act (FDMA) was amended in 1997 as part of the Food and Drug Administration's efforts to modernize its regulatory procedures. The FDMA was initially passed in 1992 to provide the FDA with the necessary resources to effectively regulate food, drugs, and cosmetic products, but it was clear that the legislation needed additional amendments in order to keep pace with changing times (Jin et al., 2020).

The 1997 amendment was written not only in response to changing threats to the health and well-being of the public, but also in response to changes in technology. For example, new methods for testing foods for contaminants or for determining whether or not a drug would work for its intended purpose became available during the 1990s. As a result, the amendment required that all manufacturers of dietary supplements, foods, drugs and cosmetics make available to FDA officials information on how they test their products, what they test them for, how they determine their safety levels and more. In addition to allowing FDA officials access to manufacturing facilities and records, the Food and Drug Modernization Act Amendment of 1997 required that manufacturers implement programs designed to ensure product quality, consistency and efficacy.

2.17 GLOBAL FOOD SAFETY INITIATIVE OF 2000

The Global Food Safety Initiative of 2000 was a landmark time in the world's history. It transformed the way we see our food and how we eat it. Prior to the GFSI, governments were the only players in the food regulation game. But companies were starting to realize that they needed to get their hands dirty, that if they wanted to succeed in an increasingly-globalized society, they'd have to take some responsibility for ensuring their products were safe and reliable.

Figure 2.15: An image showing the GFSI brief summary.

Source: https://www.google.com/imgres?imgurl=https%3A%2F%2Fimage.slideserve.com%2F76917%2Fglobal-food-safety-initiative3-l.jpg&imgrefurl=https%3A%2F%2Fwww.slideserve.com%2FGideon%2Fpresentation_76917&tbnid=LtStwRIa5PR0-M&vet=12ahUKEwi7i9y8tOb2AhU_X_EDHSVwA7UQMygDegUIARC8AQ

..i&docid=_pu9eHV2N3AA5M&w=1024&h=768&q=Global%20Food%20Safety%20Initiative%20of%202000&ved=2ahUKEwi7i9y8tOb2AhU_X_EDHSVwA7UQMygDegUIARC8AQ

The GFSI brought together stakeholders from manufacturing, retail, and foodservice, along with auditors, government authorities, and NGOs. They set out to develop a standard that would make sure that food was safe and valuable at every level of supply chain management. Their work was innovative and collaborative, and it's still being developed today. This comprehensive initiative has changed the way we interact with food forever.

2.18 FOOD SAFETY MODERNIZATION ACT SIGNED INTO LAW IN 2011

The Food Safety Modernization Act was passed in 2011, and it was the most comprehensive food safety reform legislation in 70 years. The Food Safety Modernization Act amended the Federal Food, Drug, and Cosmetic Act to provide the FDA with new powers to ensure that food is safe by expanding the FDA's capacity to prevent food-borne illness, rather than only being able to react after an outbreak has occurred. The law also grants authority to require food recalls and improved inspection procedures (Jespersen & Huffman, 2014).

It also helps clarify what foods are considered high risk, which allows for more enhanced regulation of foods that are likely to be contaminated during production. These include: shell eggs, leafy greens, produce grown in or near the ground, tomatoes, melons, sprouts, and fish. Additionally, the Food Safety Modernization Act requires all imported foods to meet the same standards as those sold domestically. It also requires importers to verify that their foreign suppliers employ good manufacturing practices (GMP) and have systems in place for hazard analysis and risk-based preventive controls (HARPC).

The FSMA is a big deal because it's the first update to food safety laws since 1938. These new rules will be strict than the previous version, so companies that produce food will have to be more careful than ever before. Some people are worried that these regulations are too strict, but they were created with two important things in mind: Your health and the health of your family. Nobody wants you to get sick from their product.

2.19 BLUEPRINT FOR A NEW ERA OF SMARTER FOOD SAFETY OF 2019

In September of 2019, the FDA released a new plan for food safety. Called "Blueprint for a New Era of Smarter Food Safety," the plan is intended to help producers create food products in the most efficient and effective way possible, increasing innovation and ensuring food safety at every step of the supply chain.

The blueprint is organized around four main goals: Build an integrated food safety system that allows all stakeholders to share information more effectively; Adopt modern, science- and risk-based preventive controls across supply chains; Improve detection and response to foodborne illness outbreaks; Develop a smarter, more rapid, and more efficient product tracing system (Jacxsens et al., 2010). The new blueprint will essentially help businesses like ours stay up-to-date with industry standards while also helping us be more innovative and serve our customers better.

CHAPTER 3

FOOD CONTAMINATION

CONTENTS

3.1 INTRODUCTION

The high concentration of chemicals contained in food causes major health hazards, therefore food contamination is a serious problem. Protecting the population from varying degrees of tainted food toxicity has become a difficult issue. The causes, kinds, and health effects of chemical contamination in food are discussed in this article. Food contamination may be caused by naturally occurring contaminants in the environment or by contaminants introduced purposefully by humans. Food contamination can also be caused through the stages of food manufacturing, packaging, transportation, and storage. These chemical pollutants have serious health consequences, ranging from mild gastroenteritis to deadly hepatic, renal, and neurological disorders.

Figure 3.1: Food contamination is a global health hazard, with one out of every ten people being unwell each year as a result of tainted food. As a consequence, businesses are required to maintain the greatest levels of food safety. Failure to verify that all food products are safe to eat could have disastrous consequences.

Source: https://www.newfoodmagazine.com/article/15069/food-contaminants-eu-contaminant-legislation/

Food contamination occurs when food is contaminated by a foreign substance, which can be chemical, biological, or physical. Food contamination happens when germs, viruses, and bacteria, are allowed to infect the food due to improper cooking, storage, or handling. If you consume tainted food, you may experience unpleasant and possibly fatal symptoms. As a result, food contamination should be prevented to the greatest extent possible. A brief explanation of the three main forms of food contamination is given below:

- Biological – Food that has been contaminated by organisms or the compounds they generate is referred to as biological contamination. Humans, bacteria, insects, and rodents all produce biological material. Viruses and bacteria are the two most prevalent forms of biological contamination, and they can cause food poisoning from Norovirus, *E. coli*, Listeria, and Norovirus, to name a few. Hand washing and sanitizing food handling equipment are two of the most effective strategies to avoid viral and bacterial contamination (Jackson, 2009).
- Physical – When a foreign object contaminates food, it is referred to as physical contamination. This can occur at any point during the manufacturing process, and it could contain plastic fragments, steel wool, or band-aids. An individual who mistakenly consumes the foreign object may be injured as a result of physical contamination. Physical contamination comes with the additional risk of biological contamination from the unwanted object.
- Chemical – Food that has been contaminated with a natural or manufactured chemical component is referred to as chemical contamination. These contaminants are especially hazardous because they expose individuals to a variety of harmful compounds, some of which are deadly. Pesticides transported from the soil in which the food is grown or throughout the manufacturing process can contaminate food at any point in the process. Chemicals must be stored separated from food in order to avoid chemical exposure.

Food may have been mismanaged in numerous ways along the food supply chain before it causes illness. Once the food has been contaminated, further mismanagement, including undercooking it or leaving it out on the counter at a temperature that is unsafe, can increase the risk of a foodborne illness. For food that is kept at room temperature, germs may multiply quickly; a small number can swiftly increase to a high number in just a few hours (Hellwig et al., 2019). Because some germs create toxins that are not eliminated by heat, boiling or reheating food after it has been left at room temperature for a long time does not always render it safe. Below we will discuss some of the common modes, types, effects and prevention of food contamination.

3.2 ENTRY OF CONTAMINANTS INTO THE FOOD CHAIN

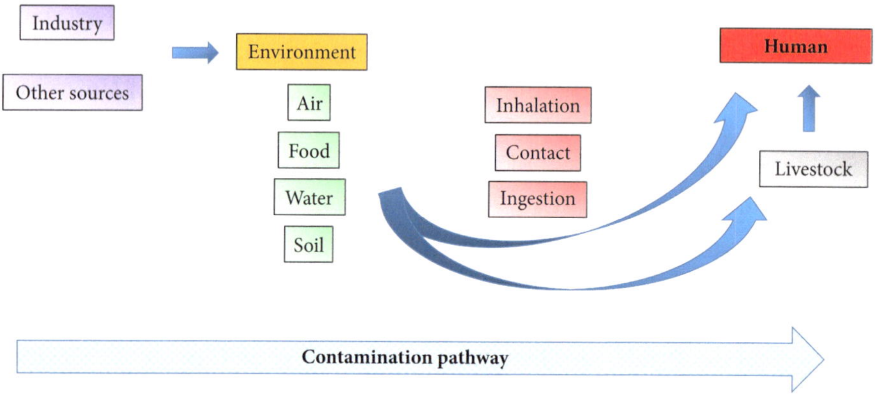

Figure 3.2: Chemical contamination is a major problem in the food industry all around the world. There are numerous potentially harmful compounds in the environment that could contaminate human food. They can be both inorganic and organic, and they can come from a variety of places.

Source: https://www.hindawi.com/journals/jt/2019/2345283/

Food is an important contributor to human health and well-being, as well as a major source of anxiety, stress, and pleasure with diseases caused by contaminated food being one of the causes of tension and worry. Food contamination can occur for a variety of causes. Food production involves a complex chain of processing steps, each of which poses a risk of chemical pollutants infiltrating the food. Transportation of food can potentially set the stage for food contamination, particularly in unsanitary conditions. Similarly, some chemicals are intentionally added during the food preparation process in order to extend the shelf life of a product. When meat is prepared in the kitchen, pollutants may be present; however, transmission is mostly dependent on the efficacy of kitchen hygiene. Chemical pollutants infiltrate the food chain naturally, as do infections found in the environment, and some crucial raw foods, such as poultry meat, have high bacterial counts (Ergönül, 2013) (Fritsche, 2018).

Pre-harvest contamination is the stage in which food is grown in the field (plant source) and animals are bred for eventual slaughter (animal source). As a result of environmental contamination, some of the chemicals are already present in food. Microorganisms, diseases, pesticides, heavy metals, and other dangerous chemicals can contaminate food if it is cultivated on contaminated soil or irrigated with contaminated water. This could also be the effect of excessive pesticide and fertilizer use.

The post-harvest stage is the final stage in the food production process. There are four additional stages in this process. Food processing is the first step following harvesting, and it entails converting plants or animals into food. For each and every food item, there are a plethora of various food processing stages. Washing or sorting, for instance, might be a food manufacturing process. Shredding, trimming, and slicing are examples of other operations. Slaughtering is the initial step in the processing of animal origin products, after which they are diced into bits. Meat and meat products, especially poultry, can be cooked, frozen, or smoked, and sausages are occasionally made by mixing them with other components like potpie.

Figure 3.3: In the United States, viruses are the most common cause of food poisoning. Bacteria are the second most common cause. Approximately 31 bacterial and viral pathogens cause over 9.4 million food poisoning infections each year, with an additional 48 million cases of nonspecific food poisoning (undiagnosed). Every year, around 128,000 individuals are hospitalized and 3,000 people die as a result of food poisoning.

Source: https://www.emedicinehealth.com/food_poisoning/article_em.htm

Cross-contamination to food can also occur if the surfaces of processing units or storage bins are contaminated. This applies to both plant and animal-derived foods. Most of the time, there's a probability that the final beef product will be infected with germs from the animal's bowels. Another example is food distribution and transportation: transporting food from a farm to a consumer's location necessitates the use of transportation and distribution. The distribution and transportation process are complicated by far-flung locations such as grocery stores, restaurants, and markets. This is an important step since food safety is in jeopardy (Fung et al., 2018).

Contaminants may indeed be found in raw foods as a result of contamination sources in the environment. Vehicle exhausts of diesel and petrol, as well as cross-contamination in the vehicle used for food delivery, are major sources of contamination during food transportation. Long-distance transportation ships are frequently contaminated by disinfectant chemicals or other sources. High barriers, which are intended to safeguard food by wrapping it during long-distance transportation, are not usually assessed for their barrier characteristics, resulting in contamination. Contaminants can enter during the cleaning phase of food production and preparation due to residues left on the surface of food handling machinery from antiseptics and cleaning agents.

Another source of pollutants is the heating treatment used in the manufacturing process. The use of high cooking temperatures in both homes and industries is a common food processing practice. Cooking at a high temperature, along with environmental circumstances, can result in the creation of hazardous chemicals that compromise food quality and safety. During food processing processes such as canning, baking, grilling, roasting, heating, hydrolysis, or fermentation, toxic chemicals such as furanes, acrylamide, chloropropanes, nitrosamines, or PAHs are generated.

In the process of preparing food, frying is a major producer of a variety of hazardous chemicals. Microwave cooking also has the potential to introduce pollutants into food, as the food is often cooked in the wrapping film or container within the microwave oven. Microwavable packaging materials include plastics, composites, and paperboard, and components of these materials can transfer from the package to the food during heating, resulting in food quality and safety issues.

From the farm or the fisheries to the dining table, there are various steps. The food production chain is what we call these steps. Contamination can happen anywhere in the supply chain, including during production, processing, distribution, and preparation. Growing the plants, we eat or rearing the animals we eat are both examples of production. The majority of food is produced on farms or ranches from domesticated animals and vegetation. Some other foods such as game, fish, and mushrooms are captured or gathered in the wild (Gill et al., 2007). Contamination can occur during production in a number of ways. For instance, if a hen's reproductive systems are sick, the yolk of an egg can be contaminated in the hen before it is laid. Vegetables and fruit can be poisoned before harvest if fields are treated with tainted water for irrigation; and in some tropical reefs, fish may absorb a toxin from the smaller sea animals they eat. In regards to processing, it refers to the transformation of plants or animals into what we identify as food. For different types of foods, processing entails various steps. Processing food might be as basic as washing and sorting or as complex as shredding, slicing, or trimming. Pasteurization is a very common method; it is also used to make cheese. Roasted, chopped, or ground nuts are all options (for example with peanut butter). Slaughter is the first step in the processing of animals. Poultry and meat can then be chopped or ground or into pieces. They can also be frozen, cooked, or smoked, and combined with other ingredients to make sausages or entrées like potpies.

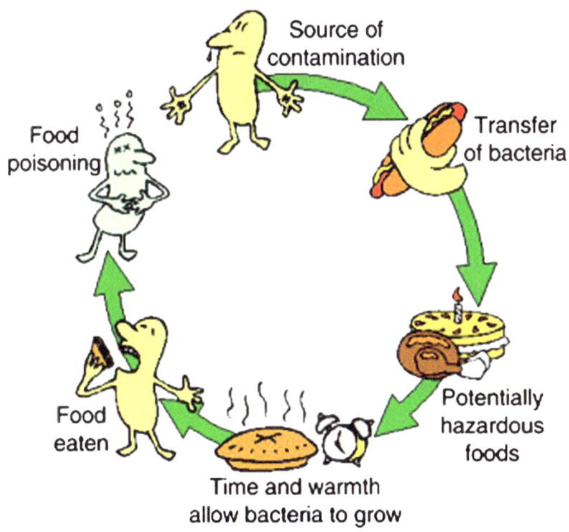

Figure 3.4: Food contamination is caused by a variety of contamination methods. Cooking and storage temperatures, improper hand washing, and animal manure are all little variables that might cause food-borne disease. Humans who did not wash their hands properly can introduce microbes into their food.

Source: https://www.sciencedirect.com/science/article/pii/B9780128115152000019

Some examples of ways in which food may be contaminated during processing include: contamination of fruits and vegetables if there is use of contaminated ice or water to wash, chill, or pack the vegetables and fruits; germs on an animal's hide coming from the intestines and other internal organs of the animal can enter into the final piece of meat during the slaughtering process; and in addition, if pathogens contaminate food processing surfaces, such as the storage bins and the processing lines, germs can spread to foods that touch those surfaces.

Food distribution refers to the movement of food from a processing plant or farm to a consumer or food service establishment such as a cafeteria, restaurant, or hospital kitchen. This phase could entail only carrying foods once, like delivering produce from a farmhouse to a local farmers' market. It could also be a multi-stage process) (Hermann et al., 2018). Frozen hamburger patties, for example, might be transported by truck from a meat processing factory to a major supplier, then held for a few days at the supplier's warehouse before being trucked again to a smaller distribution center for a chain of restaurants and eventually delivered to a single restaurant. Food contamination during distribution can occur when chilled food is left on a loading dock for an extended period of time in warm weather, allowing bacteria to proliferate; or when fresh produce is placed onto a truck that has not been cleaned after transporting animal products or animals themselves.

Preparation refers to the process of getting food ready to eat. This stage can take place in a restaurant, a house, or an institution's kitchen. It could entail following a complicated recipe with numerous ingredients, just heating and serving a dish on a plate, or by the simple act of opening a package and consuming the contents. Food is contaminated during preparation because: a food worker may stay on the job while unwell and he/she does not thoroughly wash his/her hands after using the restroom, he/she may end up transferring germs by touching food. In addition, if a cook uses a knife or a cutting board to chop raw chicken and then uses the same cutting board or knife to slice up tomatoes for a salad without washing the knife first, the tomatoes may become infected with germs from the chicken. In the refrigerator, if juices from the meat get on products that are meant to be eaten raw, contamination might occur. These are just but a few.

3.3 TYPES OF FOOD CONTAMINATION

Chemical, physical, microbial, and allergenic contaminations are the four basic types of contamination. These four species of bacteria can contaminate any food. This is why it is a legal requirement for food handlers to verify that the food they prepare is free of harmful pollutants and safe for consumption. Contamination of food is extremely harmful. Eating tainted food has a wide range of repercussions, from disease to death. As a result, it is vital that you understand the four types of contamination and how to avoid them from compromising food safety (Havinga &Verbruggen, 2017).

3.3.1 Chemical Contamination

Chemical pollutants' effects on consumer well-being and health are often only apparent after many years of low-level exposure (e.g., cancer) and processing. Chemical pollutants in foods, unlike food-borne microorganisms, are frequently unaffected by heat processing. Chemical pollutants can be characterized based on where they came from and how they got into the food.

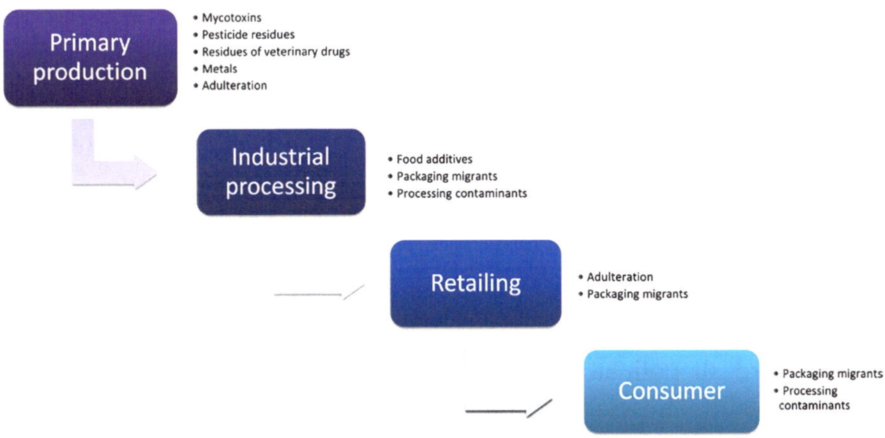

Figure 3.5: Chemical food contaminants are compounds that are not naturally present in the raw materials used in food production and are not introduced during the manufacturing process.

Source: https://www.researchgate.net/figure/Some-possible-routes-of-chemical-contamination-in-the-food-chain_fig3_316732456

The term "chemical contamination" refers to the presence of chemicals in places where they should not be, or in amounts that are higher than what is considered safe. Chemical risks are one of the most common sources of food contamination and outbreaks of foodborne disease. Chemical pollutants come from a variety of places, including soil, the environment, disinfection by-products, personal care items, water, air, and packaging materials, and they can come from anywhere from the field to the plate. Almost all mass-produced everyday use items, such as deodorants, detergents, plastics, disinfectants, and insecticides, and so on, are inhibited by chemical pollutants. Even the food we eat and the water we drink are not immune to the invasion of chemicals in dangerous concentrations (Griesche & Baeumner, 2020).

Food consumption has evolved from a simple trading channel between producer and consumer to a complicated network involving a number of different stakeholders. Drinking water, like food, can include toxins that have major health consequences not only for humans but also for marine life and other organisms that consume contaminated water. Municipal and Industrial discharges, natural geological formations, rural and urban run-off, water delivery materials and processes for drinking water treatment, are all sources of these toxins. Human operations like horizontal drilling and hydraulic fracturing have increased energy production while also increasing the risk of pollution of drinking water. Heavy metals (e.g., chromium, copper, mercury, and nickel) can pollute drinking water supplied from groundwater, resulting in a rise in cases of carcinogenic and non-carcinogenic health abnormalities, including fecal contamination.

This type of source of water contamination is particularly common in low- and middle-income nations. Pharmaceutical by-products are also harmful and have been recognized as a source of chemical contamination of water. Arsenic, pesticides, fluoride, lead, aluminum, radon, and disinfection by-products, are all pollutants found in drinking water. They have a wide range of health impacts, including cancer, cardiovascular disease, poor reproductive results, and neurological illnesses. It has also been discovered that moms who drink chemically tainted water, particularly those who are less educated, have significant effects on their infants' gestation and birth weight.

Individuals' health is seriously harmed by chemically polluted food. The negative consequences range from mild stomach issues to significant health complications. Chemical pollutants have been connected to serious consequences, a loss of personal control, and long-term ramifications. The most common way for humans to be exposed to metals is through food

(Zanin et al., 2017). Cadmium and lead are two metals that can easily infiltrate the food chain. Heavy metals can deplete particular nutrients in the body, lowering immune defenses, impairing psychosocial functions, and causing intrauterine development retardation. Consumption of heavy metals has also been linked to malnutrition and an increase in the incidence of gastrointestinal illnesses.

Contaminants in food are also a major cause of cancer. Exposure to polychlorinated biphenyls (PCBs) from contaminated food can have a negative impact on children's neurological development and immune response. Pesticides in food as pollutants have serious health consequences. Excessive quantities of these chemicals in food can induce neurological and renal damage, congenital defects, reproductive issues, and even cancer. Pesticide buildup in the body's tissues can potentially lead to metabolic degeneration. Industrial pollutants including arsenic, PCBs, and lead in both food and water also provide a risk of neurodevelopmental disorders such as mental retardation, cerebral palsy, autism, and attention deficit disorder.

Figure 3.6: At certain concentrations, chemical pollutants can be hazardous to one's health. It is vital to control their concentrations in food and decrease consumer dietary exposure. Several measures, particularly better practices and regulatory restrictions are in place to mitigate the danger from these pollutants and minimize the levels at which they are found in food.

Source: https://www.ifsqn.com/forum/index.php/gallery/image/70-chemical-contamination-control/

Chemical contamination of food has become a serious worry, posing health risks in the process. The majority of food contamination happens as a result of naturally occurring poisons and pollutants, or as a result of food processing, packing, preparation, storage, and transportation. The detection of such pollutants is becoming easier as technology develops. There are, however, a number of pollutants that are currently unknown, and study is ongoing in this area. Despite the fact that the government has taken necessary steps to decrease individual exposure to food pollutants, more has to be done to lessen the health risks and diseases associated with chemical food contamination.

3.3.2 Biological Contamination

Substances or organisms produced by organisms that present a threat to human health are referred to as biological hazards. They are a major source of foodborne illness outbreaks; hence they are a serious problem in the food processing industry. These organisms have the potential to cause intoxication, infection, and even death in humans. Infection happens when organisms infiltrate the host's body and reproduce (Xiang et al., 2019). When bacteria produce toxins that damage the body, intoxication ensues. Because bacteria are rapidly killed by heat, infection can be avoided by properly handling and processing food products. Some bacteria that create spores, on the other hand, can withstand high cooking temperatures. Steam cooking, for example, dramatically lowers non-spore-forming infections but simply inactivates spores. *Clostridium perfringes*, *Clostridium botulinum*, and *Bacillus cereus* are examples of spore-forming bacteria.

Toxins are formed when a large number of toxin-producing microorganisms are present. Before creating poisons, bacteria must first multiply in the food. Toxins are notoriously tough to get rid of. They can withstand conventional cooking temperatures as well as temperatures above 80°C (176°F). Controlling the proliferation of toxin-producing bacteria reduces the danger of poisoning. *Staphylococcus aureus, Bacillus cereus*, and *Clostridium botulinum* are among them.

Microorganism growth in foods is influenced by a number of factors, including: The ideal temperature for microbial growth varies depending on the microbe. Psychrotrophs, like *Listeria monocytogenes*, can grow at temperatures as low as 4°C (39°F), but thermotrophs may thrive at temperatures as high as 45°C (113°F). The acidity or alkalinity of a product is proportional to its ph. Bacterial growth is influenced by the pH of goods.

Most bacteria thrive in a pH range of 5 to 9. The Water Activity (aw) indicates how much water is present in the product. Bacteria will thrive better if there is more water available.

Preventing biological risks is the most effective method of control. The use of Hazard Analysis and Critical Control (HACCP) and Good Manufacturing Practices (GMPs) in your plant will aid in the prevention of biological hazards. During food manufacturing, GMPs ensure that risks to employees and the environment are minimized. Hazards that may be present in ingredients and packaging materials, as well as those that arise during food processing, storage and packaging, are all controlled by HACCP.

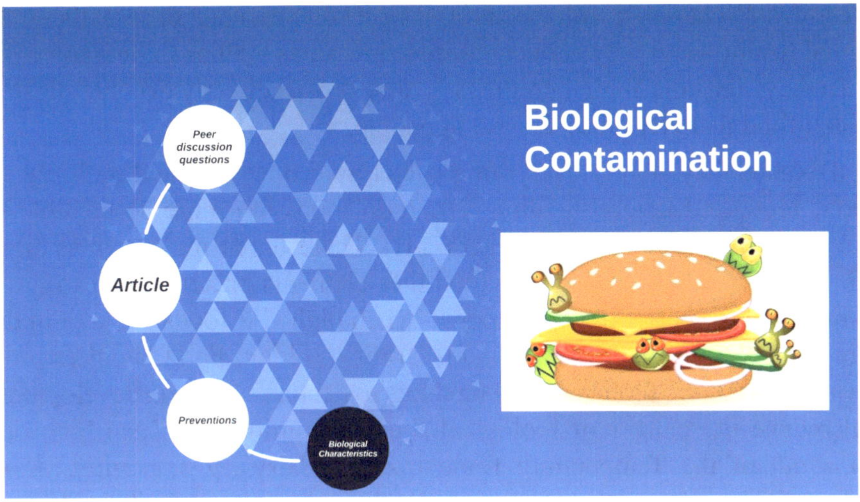

Figure 3.7: Being a sloppy person who does not take care to clean themselves and prevent food contamination is referred to as poor personal hygiene. Hand-washing after coughing, sneezing, or using a tissue is important because bacteria on your hands might get into your meals.

Source: https://prezi.com/p/0v_4ekj9txxc/biological-contamination/

Some foods are thought to have a significant microbial contamination risk. With the provision of water, food, and neutral pH, the nature of these foods creates an ideal habitat for bacteria to grow. Here are a few examples: Seafood, rice, poultry, eggs (both cooked and uncooked), dairy products and dairy goods, meat, unpasteurized juices, and vegetables and fruits that have been prepared.

The danger zone is defined as any temperature between 5 and 63 degrees Celsius, with food stored at these temperatures being particularly susceptible to microbial contamination. Bacteria quadruple their numbers every 20 minutes in such optimal conditions. This means that in just seven hours, a single bacterial cell can multiply to 2 million (Wilcock et al., 2004).

A number of pollutants, particularly viral and bacterial diseases, can pose a biological threat. Mycotoxins, which are the chemical contaminants produced by these organisms, are described under fungal pollution. Biological hazards that are frequently monitored under the Animal Food Contaminants program include *Listeria monocytogenes*, Salmonella, and infectious Escherichia coli bacteria. These viruses may pose a threat to animal health when they eat contaminated food, as well as to human health when they consume animal products or are exposed directly to contaminated animal food that can cause disease in people.

Because they supply everything bacteria need to exist and multiply — water, food, and neutral acidity — certain foods are more susceptible to biological contamination than others (pH). These foods are known as high-risk foods. Food Handlers offer the other elements bacteria need to grow — time and the correct temperature — when high-risk foods are kept in the Temperature Danger Zone (5°C – 60°C) for too long. Food handlers must always follow safe food handling methods to decrease bacteria development and reduce the danger of biological food contamination: Keep high-risk items out of the Temperature Danger Zone (dairy, eggs, poultry, meat, and shellfish); properly purchase, store, defrost, prepare, cook, and serve high-risk meals (Wertheim-Heck et al., 2019).

Contamination may also come from the worker if proper hygiene is not observed. Water contaminated with bacteria such as fecal coliforms, *E. coli*, heavy metals such as arsenic, lead, and other contaminants used for irrigation and pesticide application can cause foodborne illnesses. Furthermore, animals bred in animal homes with contaminated water get infections, which are transmitted to humans when these animals are consumed as food. Because of their high-water holding capacity, crops like carrots and lettuce are sensitive to *E. coli* and salmonella.

Figure 3.8: Gastroenteritis is usually contracted by ingestion of tainted drink or food, or through direct contact with a person, surface, or object - all of which are caused by a lack of sanitation and/or hygiene. Vomiting and watery diarrhea are the most common symptoms, as are stomach discomfort, cramps, headache, nausea, and fever. Individuals might get dehydrated as a result of diarrhea and vomiting.

Source: http://foodsafetytrainingcourses.com/blog/articles/food-safety-news/introduction-to-biological-food-hazards/

Feed lots, transportation, and lairages, among other things, can contaminate meat and poultry during the pre-slaughter and slaughter stages. Meat and poultry can be contaminated with bacteria that cause food poisoning if they are fed infected food. Other materials such as clothing, footwear, and insects, rodents, pets, animal and human feces can all contribute to meat and poultry contamination. Cross-contamination can occur when contaminated equipment, like as a butcher knife, is utilized. Because the contamination of non-aquaculture-grown fish is unclear, it poses a significant risk to human health. The quality of the fish in aquaculture is affected by the water environment (Wahyuni et al., 2019).

Cross-contamination may occur as a result of additives added during the packaging of food. Food spoils when it is not kept at a safe temperature during transportation and distribution. Dry foods such as flour, rice, cereals, and others are packaged with paper and board, which can lead to cross-contamination. Cross-contamination occurs when edible things are carried with animals or in a truck that was previously used to carry cattle (without disinfection and cleaning). Any biological hazard that occurs at this stage will have a detrimental influence on meat and meat products. The last word is Food preparation refers to the process of cooking food so that it is ready to eat.

Food preparation can take place at a restaurant, a school, an institution, a workplace, or home kitchen, among other places. The phase entails mixing raw food with ingredients and spices, as well as additional steps such as heating, boiling, baking, and serving. Contamination can come from the cook's hands (if gloves aren't worn), as well as equipment (that has not been washed or sterilized) used in the kitchen, such as knives, chopping boards, spoons, and other utensils. Food of both plant and animal origin is polluted in the same way during food preparation (Tirado et al., 2010). Chefs should maintain good hygiene and sanitation, as well as cleaning and washing of equipment and surfaces on a regular basis. Contamination can also be caused by the storage conditions of prepared foods, regardless of whether they are kept in the freezer or refrigerator. When raw meat is placed with raw goods, the juices from the meat might leak into the raw food, producing cross-contamination, which is especially dangerous when the food is intended to be consumed raw.

3.3.3 Physical Contamination

When a hazard in the form of a physical item enters food, it is referred to as physical contamination. It can be highly upsetting and stressful to discover physical contamination in your meal. It may also cause sickness because some physical food pollutants serve as a breeding environment for bacteria. Choking hazards can also be caused by physical pollutants. The great majority of the time, physical food pollutants enter meals by mistake. They are occasionally placed there on purpose as a form of direct sabotage. A food incident occurs when the quality or safety of a food product is put into question (Todt et al., 2009).

Certain items may be recalled as a result of a food incident. Catering businesses must be able to identify their supply chain in this situation so that they may take the appropriate measures and return or dispose of any compromised products. Food that has been contaminated should be removed from circulation, and the local authority where the business is located should be notified. The FSA Incidents Response Team of the Food Standards Agency can help with recall guidance. According to government data, more than 25 food products had to be withdrawn from circulation due to safety concerns in the last six months of 2019. Businesses are urged to do a Root Cause Analysis after a food-related occurrence (RCA). The RCA's findings can assist firms in improving their operations and preventing a repeat catastrophe. They can also assist food companies in learning more about traceability and how to keep track of their supply chain.

Figure 3.9: Stones, bones, plastic, hair, jewelry, insect corpses, and fabric are examples of physical pollutants detected in food. Furthermore, if there are issues with the food premises or equipment, such as peeling paint or loose screws in a piece of equipment, they might go into the food.

Source: https://slideplayer.com/slide/17011683/

Food can be contaminated by a variety of causes. It could be because something dropped into the dish from the person who made or served the food, a piece of packaging was left behind, or insufficient cooking and serving tools were utilized. A physical contaminant's presence could be an unintended consequence, such as a strand of hair from a waiter slipping into a dish on its route from the kitchen to the customer's table. It could also be a sign of a more significant issue within the company, such as a disorganized and filthy food processing area.

When you open a can of peaches or a bag of cereal, you'll usually find exactly what you're looking for: peaches (with syrup) in the can and cereal in the bag. But have you ever discovered an unexpected item in your meal, such as a bit of plastic in your peaches or a piece of metal in your cereal bag? Physical contamination of food is when you find strange items in your food. One of the three types of food contamination is physical contamination. There is both chemical contamination and biological contamination. Pesticides, medicines, cleaning agents, and allergies are all included in the chemical category. Salmonella, listeria, and other bacteria are examples of biological contamination (Sinha, 2009).

Physical contamination can occur in the natural world. This can involve things like insects getting into your broccoli or bones showing up in supposedly boneless seafood. Food makers strive to eradicate all physical pollutants, but it is sometimes impossible to do so completely. We can't stop every insect from concluding that broccoli seems like a good home in the fields where fruits and vegetables are cultivated. This is especially true with the growing trend of organic foods, where the lack of pesticides makes it difficult to guarantee that the broccoli doesn't contain insects. If you buy broccoli from a farm, there's a strong possibility you'll find an insect or two in your purchase.

Producers usually wash their produce before selling it in order to remove even more insects. Insects, on the other hand, can elude all measures at eradication and end up on the table of the consumer. Insects in food aren't usually considered a health risk unless they carry diseases that are harmful to humans. Insects, in particular, are a quality issue since consumers dislike finding them in their meals.

A bit of plastic from a bag of sugar falling off and getting mixed into a cake is an example of accidental physical contamination. A metal shaving from equipment breaking off and ending up in the macaroni is another example. Food manufacturers use a number of measures to guarantee that this does not happen because some of these issues can be toxic to consumers. Many food producers, for example, demand that all products be run through a metal detector to detect any metal contamination.

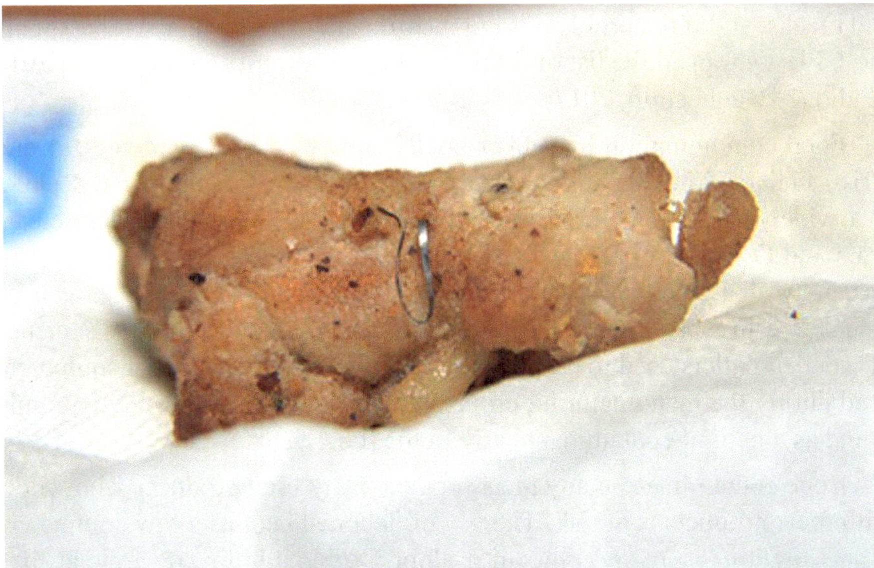

Figure 3.10: Foodborne physical pollutants are a global health and safety problem. There have been high-profile examples of foreign bodies like glass, plastic, and metal entering the food chain in significant markets such as the UK and the US. Even while such instances, whether accidental or purposeful, are uncommon, they can pose a significant health risk to consumers.

Source: https://www.foodonline.com/doc/ensuring-food-safety-through-the-prevention-of-physical-contamination-0001

Metal can end up in food during the farming process as well as during the many stages of food preparation. If a metal blade cracks when cutting or harvesting plants, metal shards from the blade could end up in the food inadvertently. A machine may lose a nut or bolt that falls into the food during processing, or a defective piece of equipment may produce metal shards and shavings that end up in the meal.

When glass containers are used in the manufacturing process, there is a chance that some glass will wind up in the final product. Glass contamination can occur through any form of packaging method; however, it is more likely when the glass is handled mechanically. As a result of overhead lights shattering in a production plant, glass shards can sometimes seep into the food. When hard plastic equipment and tools (sieves, buckets, paddles, etc.) are used in facilities, the plastic might wear out over time. This causes cracks and breaks, and if the problem is not identified, the plastic may end up in the food. Plastic topped the list of foreign pollutants discovered in food in 2018 (Sadiq & Beauchemin, 2017).

Food contamination by rubber goods grew by 22% between 2015 and 2016, underscoring the importance of food product monitoring. Rubber is a widely utilized material in manufacturing, where it is used to create seals and O-rings to prevent food leakage. When they wear out, however, they may shatter into shards and fall into the food. Wood pieces can be discovered in a wide range of foods; they are frequently the consequence of wooden pallets used to carry the food or wooden utensils and equipment used during the manufacturing process. Stones and boulders are commonly found as a natural contaminant early in the food chain.

Bone contaminates many meat products; any food producer who works with meat products is at risk. The use of deboned beef as a raw component is one solution that many companies adopt. During grinding operations, they may additionally employ a specialized bone removal device. Detection is typically the most difficult aspect of dealing with food poisoning. Physical inspections are ineffective in detecting foreign materials embedded in food, which is why X-ray inspections are becoming increasingly popular among food manufacturers.

Many physical food pollutants can be avoided by modifying operations within a facility, according to the FDA. Even the most severe procedures, however, can go wrong at times. Machines will fail, gaskets will fail, cutlery will fracture, and jars will shatter. There's no guarantee that no matter how many protocols and protections put in place, they'll work 100 percent of the time (Smith & Pillai, 2004). FlexXray has become the firm that food producers trust to verify that their product is free of debris and physical contamination, thanks to cutting-edge equipment that can identify glass, gasket, plastic, bone, metal and other pollutants as small as 0.2mm (or even smaller in some situations).

3.3.4 Allergenic Food Contamination

Cooking for the family is a delightful everyday chore that can bring the family closer together. However, when you are cooking for someone who has food allergies, things get a little more complicated. This is the situation. When you have an allergy, the same goes for eating out. Knowing how to handle cross-cultural situations is essential. It is critical for your health to be aware of the possibility of contamination and to take the required safeguards. Then one food comes into contact with another, it is referred to as cross-contamination. Food causes their proteins (which cause an allergic reaction in your body) to combine. For instance, utilizing a barbecue to first fry eggs, then bacon, and then only bacon. In between, wipe off the barbecue with a moist cloth. Because this method of cleaning is insufficient to entirely remove all trace amounts of egg left on the barbecue, little traces of egg may end up on the bacon. If someone who is allergic to eggs eats this bacon, their allergic reaction will be the same as if they had eaten an egg.

Figure 3.11: According to some studies, up to 8% of persons in the United States have food allergies, with around 40% of them having a history of severe responses needing emergency treatment. Hives, a runny nose, and/or trouble breathing are the most common symptoms. Some signs are so severe that doctors must use intravenous drugs, inhaled drugs, and IV fluids to keep the patient alive.

Source: https://www.dininggrades.com/blog/allergic-reactions-to-food-can-kill/

When a food allergen contaminates a naturally allergen-free food, this is known as cross-contamination. Bread crumbs from a typical slice of toast, for example, maybe left behind in a toaster during food preparation. These crumbs in the toaster could contaminate a wheat-free slice of bread, producing an allergic reaction in people who have celiac disease or gluten sensitivity. This issue can also arise in manufacturing facilities. Food goods containing tree nuts, for example, can contaminate the processing of other foods produced on the same line. As a result, if a product's manufacturing site also produces foods containing one of the top food allergens on the same processing equipment, a warning must be included.

Cross-contamination can happen practically anywhere, including at home, at restaurants, at school, and on production lines. Cross-contamination occurs during the preparation, cooking, storage, and even serving of food. It might be as simple as cooking gluten-free pasta in the same pot as regular pasta, eating a salad with a bit of cheese cut from the top, or using a knife that hasn't been properly cleaned from the peanut butter jar. These conditions can cause life-threatening reactions in people with food allergies. Follow these tips to help protect your food and lessen the likelihood of an adverse response (Sargeant et al., 2009).

Food manufacturers frequently process and package allergen-containing items in the same facility or equipment. To protect yourself, follow these steps: Read the food labels carefully for any mention of the food to which your allergy; Keep an eye out for comments like "processed in a plant that also manufactures wheat" or other similar remarks. These can alert you to the possibility of nuts or other common food allergies being present. It is always better to contact the company and ask about your worries about safe practices if you are unsure about the information or do not see it on the label.

In most households, members of the family with food intolerances avoid bringing foods containing those allergens into the house because of the high risk of anaphylaxis. This is the most secure method of avoiding cross-contamination. All members of the family can live in a healthier, allergy-free environment if these foods are removed from the house. If you must keep foods containing these allergens in your home, follow these guidelines: Foods containing allergies should be kept away from common food preparation and serving locations; Foods containing allergies should also be consumed in specific sections of the kitchen to prevent residue from spreading throughout the house. Everything should be properly labeled; Keep in mind that if babysitters, grandparents, or friends are in charge of the food or dining at your home, they must be able to understand the dangers of the meals; and clean all food surfaces before and after food preparation, and designate particular food preparation locations and utensils for "allergen-free" use.

Figure 3.12: This issue can also arise in industrial plants. Food goods containing tree nuts, for example, can compromise the preparation of other foods produced on the same line. As a result, if a product's manufacturing site also produces foods containing one of the major common allergens on the same production equipment, a warning must be included.

Source: https://www.verywellhealth.com/how-to-avoid-cross-contamination-with-food-allergies-1324003

When choosing a restaurant, check to see if the staff is familiar with food allergies and has a food safety strategy in place. Before placing an order, speak with a manager to ensure that the staff is aware of your food allergy requirements. Despite the greatest efforts of a restaurant, cross-contamination can still occur if all proper standards are not followed. In restaurants, the most typical sources of cross-contamination are frying oil, griddles or grills, and woks. In a restaurant, the actual cooking of food poses a greater risk than conventional prep sections, which are typically better observed and understood.

When dining out, do not be hesitant to double-check the food preparation with your server. Keep in mind that fried foods are frequently a source of cross-contamination, as heating them in the same oil can result in an allergic reaction. If your food order is incorrect and you must return it, make sure the server understands that you require a new meal. Removing the inappropriate item, such as an egg or a piece of cheese that may have come into contact with the rest of the meal, is not allowed because the meal has been cross-contaminated. Salad bars, buffet-style restaurants, and ice cream parlors should all be avoided (Redmond & Griffith, 2004).

3.4 FOOD CONTAMINATION PRECAUTIONS

Precautions are required for the prevention of a variety of diseases and food losses. Losses are more expensive than prevention, and an economic examination of food safety expenses revealed that investing in preventive is much less expensive than investing in an after-event. Proper sterile situations should be preserved for manufacture and storage, proper cooking should be done for raw foods, sterile packaging should be followed for plastic-wrapped ready-to-eat food items, adequate temperature and humidity should be maintained in the working environment of food, and HVAC should be used to ensure food safety.

Unused food products, for example, must be stored in refrigerators at 4°C-8°C, and frozen meats and seafood must be stored in deep freezers, among other things. Dangers such as biological, chemical, and physical hazards can all be considered while taking measures. Elimination systems, such as metal detectors and X-ray equipment in the case of animal meals, can be used to eliminate physical dangers. Similarly, in terms of chemical dangers, eliminating the use of fertilizers, pesticides, insecticides, and other chemicals in the agricultural field can reduce food contamination, all of which are significant environmental pollutants, and food contamination begins with meat, eggs, and milk.

Figure 3.13: When the interior temperature of food reaches a temperature high enough to kill microorganisms that can make you sick, it is considered safe to eat. A food thermometer is the only method to know if food is safe to eat. By looking at the texture and color of the food, you cannot determine if it's safe to eat (except for seafood).

Source: https://www.cdc.gov/foodsafety/keep-food-safe.html

Heavy metals, poisons, trash, and other pollutants are released by industry, polluting fields and water bodies and posing a risk of food contamination. To combat this, strict adherence to food safety rules is required. Biological risks are extremely difficult to prevent since they can arise at any point in the food chain, from farm to plate. Temperature, pH, humidity, and moisture must all be controlled to prevent microbial development, and ionizing radiations are occasionally utilized in the case of animal feed.

3.5 MEASURES OF PREVENTION

There are several laws and regulatory agencies in place to regulate chemical levels in food. Adulteration of food and the use of harmful substances are prohibited by law. As a result, robust response and surveillance systems are required for chemical hazard prevention. The FDA sets the limitations on which chemicals, such as pesticide concentrations, are authorized to be added to food. However, there are countries that rely heavily on agriculture, with pesticides seeping into groundwater and excessive quantities of food and water. Although the government has taken necessary actions to prevent individual exposure to food pollutants, there are still numerous measures that need to be implemented to reduce illnesses and health concerns (Rendueles et al., 2011).

To eliminate infections during preharvest and preslaughter, GAP (Good Agricultural Practices) must be followed during crop growth and harvesting. To prevent contamination at the preharvest stage, methods such as microbiological testing of the farm and water, fencing to keep animals out, manure application, and frequent cleaning can be done. Antibodies should be measured in animal-derived foods to determine the infection status. The more industrialized slaughtering practices get, the greater the risk of illness. Meat animals and carcasses, on the other hand, are quickly infected with germs. As a result, improved slaughter hygiene is required to avoid contamination.

Figure 3.14: When thawing goods, do not leave them out on the countertop at room temperature to avoid contamination, since germs may develop fast in these settings. Thawing food on the countertop also increases the danger of contamination, perhaps putting other items at risk in the future.

Source: https://www.active.com/nutrition/articles/8-tips-to-avoid-food-contamination

Ways in which you can prevent food contamination and food borne illnesses include:

- Handwashing – Infected persons who touch food without completely washing their hands can spread pathogens into the meal. These germs are thus conveyed to meals via trace quantities of feces on the hands. Proper hand washing techniques should be employed to prevent this mode of food contamination.
- Cross-contamination – Raw food items have the potential to contaminate food, kitchen utensils, and surfaces (i.e., meat and poultry). By using the same knife, cutting board, or other instruments repeatedly without wiping the surface or item between usage, microbes can be transmitted from one food to another. If a completely cooked item comes into contact with additional raw foods or drippings from raw foods that contain bacteria, it might become re-contaminated. These should be taken very seriously

and avoided while cooking to avoid food contamination (Qiao et al., 2018).

- Cooking and storage temperatures – Before enough pathogens are present in food to cause disease, they must grow to a bigger number. Refrigeration or freezing, in general, inhibits the growth of practically all microorganisms. Parasites, viruses, and the majority of bacteria are destroyed when food is cooked sufficiently.
- Animal waste - In healthy animals bred for food, several foodborne bacteria are present. Small quantities of intestinal fluids can infect poultry and meat during slaughter. If fresh vegetables and fruits are washed in water tainted by human sewage or animal dung, they may get infected.

To guarantee that any food contaminants that may be present, introduced, proliferate, or survive are detected during the risk assessment, it is critical to understand your whole supply chain from farm to fork. This data must also be incorporated into HACCP and policy research programs. As a basis, a clear and accurate map of each process phase and their interrelationship is beneficial. The mapping may also be used as a communication tool for stakeholders along the supply chain, as it is simple to understand and share. Poor communication, as in many situations, is the cause of food safety accidents across the supply chain.

Once the initial supply chain mapping is complete, it's critical to set up a procedure to guarantee that any internal and/or external modifications are examined to see how they affect the risk assessment. In the vast majority of situations, co-product streams and/or rework will be created, which must be evaluated as well. Contaminants may be concentrated in co-product streams in specific instances. Co-products may be destined for various market sectors than the mainstream food product, further complicating the process. In the vast majority of circumstances, these market segments will have distinct food safety, quality, and regulatory requirements.

The following is a list of some tools that may be useful in supporting effective communication with stakeholders, developing a solid understanding of the supply chain, and ensuring that events affecting food safety are flagged, allowing them to be assessed and addressed in food safety programs as needed:

- Coaching
- Evaluation and selection of suppliers

- Vendor evaluation
- Communication with suppliers
- Specifications
- Contracts
- Creating a solid relationship with customers and consumers
- Understanding how to utilize and apply the completed product.

As our global supply chain changes and research progresses, there is an increasing need for continuous risk assessment to limit the danger of contamination in foods. It is critical to assess all potential hazards throughout the supply chain using Codex HACCP principles and a thorough understanding of consumer behavior, rather than focusing solely on EU regulation, which currently covers chemical and microbiological hazards, or making assumptions about how products should be stored and handled in the household (Piggott & Marsh, 2004).

Figure 3.15: To prevent germs from spreading to raw fruit or other things, use separate cutting boards for vegetables and meat. Before and after prep, be sure to clean the countertops, cutting boards, and any other associated tools with hot, soapy water. Raw produce can also carry germs, so clean all vegetables and fruit before cutting or serving, according to Tribble.

Source: https://healthblog.uofmhealth.org/childrens-health/how-to-prevent-your-kids-from-getting-food-poisoning

As analytical tools advance, we will be able to identify new risks and detect existing ones at lower levels. For relevant results, we must use suitable sampling and analytic approaches while doing analysis. We must also remember that the possibility of economic adulteration has always existed; but, in today's global marketplace, the significance of such events is much larger. As a result, we must create and assess our food safety systems on a constant basis, utilizing a risk-based scientific approach and stakeholder engagement across the supply chain (Ambrus, 2016). Make full use of the wide range of external materials and industry guidance available, such as publications from FSA, EFSA, Codex, and industry organizations. As well as keeping an eye on external market occurrences that may prompt further assessments beyond those already scheduled as a result of structural changes.

CHAPTER 4

KEYS TO SAFER FOOD

CONTENTS

Food safety (or food hygiene) is a scientific method or discipline that describes how to handle, prepare, and store food in order to avoid food-borne illness. A food-borne disease outbreak occurs when two or more instances of a similar illness arise as a consequence of the consumption of a common food. This encompasses a series of practices that must be followed in order to prevent causing health risks. As a result, food safety frequently falls in line with food defense to protect consumers. This train of reasoning has two tracks: safety in both industry and market, and then safety for both market and consumer (Pressman et al., 2020). Food safety aspects include the origins of food, including practices related to food branding, hygiene practices, chemical additives, and pesticide impurities, and also biotechnology and food rules and procedures for the management of government agencies import and export testing and certification mechanisms for foods. When it comes to market-to-consumer practices, the common assumption is that food should be safe in the supply chain, and the greatest problem is safe delivery and prepping of the meals for the consumer.

Figure 4.1: Based on information supplied by WHO, Khmer Times business reporter Sok Chan has identified five keys to safer food preparation.

Source: https://www.khmertimeskh.com/82378/world-health-organisations-five-keys-safer-food/

Food can convey pathogens, which can cause illness or death in humans or other animals. Microbes, viruses, mold, and fungus are the most common pathogens. Pathogens can use food as a development and reproduction medium. Food preparation standards are intricate in developed countries, whereas less developed countries have fewer requirements and less regulation of those standards. Nonetheless, in the United States in 1999, foodborne pathogens were responsible for 5,000 deaths per year. Another major issue is the lack of acceptable safe water, which itself is generally a crucial factor in the spread of maladies. Food poisoning is, in principle, completely avoidable.

Even so, considering the number of people engaged in the supply chain, and also the reality that pathogens can be presented in foods regardless of how many safety measures are held, it can be accomplished (Pollard et al., 2016).

It is critical to understand and follow hygiene rules when handling food, as this ensures food safety and prevents foodborne diseases. Almost all of the time, the food handler is also the one who acts as the carrier of transmission for these pathogens, so we must recognize the crucial role we take part in prevention.

Food handler training is required and will permit you to acquire a certificate accrediting you as a food handler, enabling you to operate in the food industry. Although it is not required, it is recommended that the certificate be renewed every four years and be revised per the job. The business should then provide industry-specific training.

A food handler is someone who has direct interaction with food all through any of its stages until it gets to the final consumer as part of their job. Preparation, production, processing, wrapping, storage, transshipment, distribution, sale, supply, and service are all included.

The World Health Organization defines health as a comprehensive state of physical, mental, and social well-being, not only the lack of physical unpleasantness or illness.

So, what exactly is food hygiene? It is the collection of measures required to guarantee food safety from field to fork, or from the time food is procured until it reaches the final consumer (Pappalardo et al., 2017).

As a result, we identify food-borne illness as any illness caused by eating tainted food.

It is critical to ensure good food hygiene because it will result not only in safer food handling, but in a good company reputation, greater customer satisfaction, and the avoidance of potential health-related penalties.

4.1 FOOD CONTAMINATION. TYPES OF CONTAMINANTS AND HAZARDS

What exactly is food contamination? A contaminant, also known as a food hazard, is any foreign agent in food that has the potential to harm the consumer's health.

Dangers are categorized into three types based on their origin:

- Physics: bones, crystals, personal effects, and so forth.
- Chemicals: toxin-containing substances, cleaning products, and so on.
- Biological: living things like insects, rodents, birds, microorganisms, and so on.

The following are the primary sources of food contamination:

- The surroundings

Pests are living beings.

- Unsanitary utensils and premises
- Garbage.
- The food handler: due to a lack of hygiene.
- Other foods: This is critical because we are discussing the transmission of pathogens from one food to another. It can be administered by combining cooked and raw foods or by using the same cookware for various foods.

Food is the main defining factor of population health, nutritional condition, and productivity. As a result, it is critical that the food we eat be nutritious and safe. A variety of foodborne diseases can be caused by tainted food. You may have read in the media about health issues caused by tainted or denatured food. Food-borne disease is a major global health concern around the world. According to the National Family Health Survey in India, more than 2 lakh children under the age of five continued to suffer from acute diarrhea (Nesbitt et al., 2014). Food-borne diseases can not only cause death, but it can also harm trade and tourism, cause loss of earnings, joblessness, and litigation, and thus stymie economic growth. As a result, food safety and adequacy have become increasingly important and gained worldwide significance.

Figure 4.2: Food safety tips during the COVID-19 pandemic.

Source: https://fscluster.org/iraq/document/covid-19-food-safety-advice

Food safety and quality are essential at home, but they are crucial in large-scale food manufacturing, as well as in restaurants where meals would be freshly prepared and provided. Many foods were once processed at home. Advances in technology and preparation, greater per capita incomes and buying power, as well as rising consumer demand, have resulted in the production of a wide range of processed foods and food for health / functional foods. The safety of these foods must be determined. The quality of food, both raw and processed, is a public health concern that must be addressed. Over the last decade, the global safety challenges have shifted dramatically, and issues concerning food quality and safety have grown in importance (Meng & Doyle, 2002).

A number of factors are responsible for this:

- As habits and eating behavior patterns change, more individuals are eating out. Food contamination is more likely in commercial settings because foods are planned in bulk and managed by a large number of people. Furthermore, food items are prepared many hours ahead of time and may spoil if not properly stored.
- Furthermore, food items are primed many hours ahead of time and may deteriorate if not properly stored. The security of these foods is critical.

- Spices and condiments, as well as oilseeds, were once produced at home, and purity was not a major worry. Independent seasonings, condiments, spice powders, and blends in pre - packaged form are popular in today's world, particularly in cities and urban centers. The quality of raw foods, in addition to processed foods, is a public health issue that must be acknowledged.
- The logistics controlling bulk food transport are complex, and there is a long time lag between handling and consumption. Thus, risk analysis and safety management are critical all through mass manufacture and supply.
- Microbial adjustments, resistance to antibiotics, changed human sensitivity, and international travel have all contributed to an increase in food-borne microbial disease incidence. During the last 25-30 years, approximately half of all recognized foodborne pathogens have been discovered. There are many other food-borne illnesses with unknown causes (Mensah & Julien, 2011). This is a public health issue worldwide, and there is an urgent need to detect, identify, and recognize emerging pathogens, as well as to create ongoing monitoring systems on a global scale.
- The majority of states have signed the World Trade Organization (WTO) non-tariff contract, which has given all countries greater access to global markets and prospects to engage in international trade. In this situation, it is critical for each country to protect food safety and quality while also ensuring that foreign - made foods are of high quality and safe for consumption. Effective food standards and management measures are necessary to protect domestic food production while also facilitating trade with other countries. All food producers are mandated to meet certain quality and safety standards, as well as have their products tested on a regular basis.
- Pollution in the environment, soil, and water, as well as pesticide use in agriculture, all contribute contaminants. The use of admixtures like preservatives, colorants, flavoring agents, and other compounds such as stabilizers also necessitates food analysis for numerous elements nutrients and contaminants.

Because of the aforementioned factors, there is an increasing issue in a rapidly evolving food corporate environment for safe, healthy, and nutritious foods, which has significantly increased the context and enhanced job prospects in this sector. Before delving into the range of career paths available in this line of work, we should first understand the fundamental concepts of quality of food, food safety, risk evaluation, food safety standards, and quality management systems.

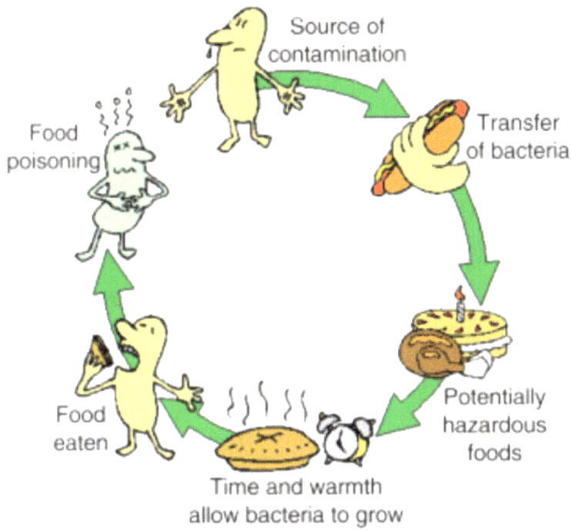

Figure 4.3: Food Products and Food Contamination.

Source: https://www.sciencedirect.com/science/article/pii/B9780128115152000019

Food-borne diseases are major public health issues in both developed and emerging nations. According to the World Health Organization, up to 30% of the population in advanced nations suffers from food-borne illnesses every year, while up to 2 million deaths are projected in emerging regions. Every day, millions around the world become ill as a result of the food they consume. This illness is known as food poisoning and is induced by unsafe microbes and/or toxic chemicals. Every year, millions of people become ill and multitudes die as a result of consuming contaminated or poorly handled foods. Food handlers with substandard personal hygiene who work in food establishments could be a source of infection for a variety of intestinal helminthes, protozoa, and microbial pathogens (McLaughlin et al., 1999).

Everyone who performs in a food and beverage organization and comes into contact with food or any tools or utensils that are almost certain to come into contact with food, like silverware, plates, bowls, or cutting boards, is considered a food handler. Infected food handlers are a major source of food-borne diseases in developed countries. Consumption of contaminated food can cause mild to serious illness, hospitalization, or even death. Diseases with brief incubation time are more likely to be recognized and credited to contaminated food than those with lengthier incubation periods, where the person may not relate their illness with the consumption of contaminated foods.

Poverty is the root cause of unsafe food intake in Africa. The absence of access to safe drinking water, poor government structural arrangements, infectious diseases, trade tension, and inconvenient environmental factors are all significant factors. The high rate of diarrheal illnesses is an indication of the African region's food hygiene situation. Food handling procedures are influenced by a variety of factors. A study conducted in Ankara, Turkey, Mekelle, and Bahir Dar, Ethiopia, found that knowledge of food handling is directly proportional to food handling practices, while a study conducted in central India, Bangladesh, and Nigeria found that handling food practices were related to food handlers' educational status (MacLaurin, 2001). Furthermore, a 2009 study conducted in Nigeria and Kenya found that the type of assumption, unsanitary equipment, and work responsibility were all factors influencing food handling practices.

Figure 4.4: Food safety is a critical matter for the people working in the hotel industry.

Source: http://buffet-go.com/food-handlers-certificate-what-it-is-who-needs-it/

Gender was also found to be related to the food handling practices of street food vendors in Nairobi, Kenya. In addition to socio demographic factors, conditions such as temperature, waste management storage, solid waste disposal, latrine condition, and food and beverage establishment hand washing facilities were linked to food handling practices. Food borne diseases are prevalent in developing countries such as Ethiopia due to poor food handling and sanitation processes, insufficient food safety laws, weak institutional arrangements, financial constraints to spend on safer equipment, and poor education for food handlers (Mou & Lin, 2014).

Considering Dangila town as a geographic training center for multiple agencies in North Gojjam, as well as the town's count of food and drink organizations, it is preferable to choose town as a survey area. Furthermore, no research has been conducted in this area to evaluate the handling of food of food handlers in this town's food and beverage establishments.

Food contamination occurs when foods become contaminated with another substance. It can happen during the manufacturing, transportation, packaging, storage, sales, and cooking processes. Physical, chemical, or biological contamination can occur.

4.2 PHYSICAL CONTAMINATION

Physical contaminants (or 'foreign bodies') include objects like hair, plant stalks, and metal and plastic fragments. A physical contaminant occurs when a foreign element enters food. If the foreign objects are bacteria, there will be both physical and biological contamination.

Hair, glass or metal, pests, jewelry, dirt, and fingernails are frequent causes of physical contamination.

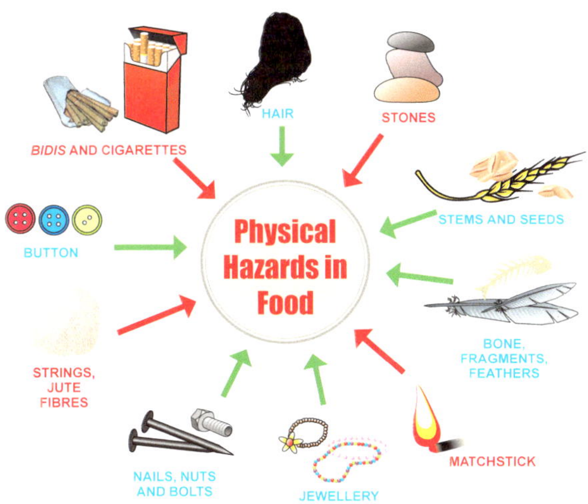

Figure 4.5: Physical hazards in foods.

Source: https://ncert.nic.in/textbook/pdf/lehe106.pdf

4.3 CHEMICAL CONTAMINATION

When food is tainted with an organic or inorganic chemical substance, this is referred to as chemical contamination. Pesticides, herbicides, veterinary drugs, contamination from ecological references (water, air, or soil pollution), cross-contamination all through food preparation, migration from packaging, the amount of natural toxins, or the use of unauthorized food additives and additives are all common sources of chemical contamination (Lytton, 2017).

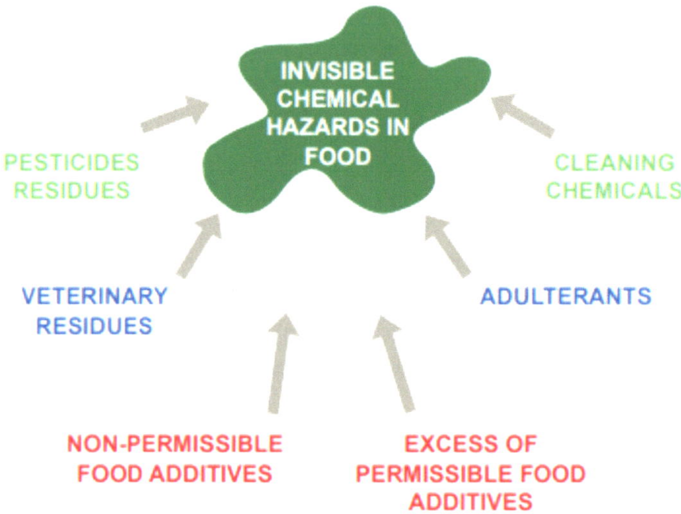

Figure 4.6: Chemical hazards in foods.

Source: https://ncert.nic.in/textbook/pdf/lehe106.pdf

4.4 BIOLOGICAL CONTAMINATION

It occurs when food is tainted by compounds produced by living organisms like humans, rodents, pests, or microbes. This would include microbial contamination, virulent contamination, or parasitic organism contamination transmitted via saliva, pest droppings, blood, or feces. Microbial infection is the leading contributor to food poisoning around the world. Bacteria seem to be likely to survive in an eco-system that is high in starch or protein, water, oxygen, has a balanced pH level, and holds a temperature between 5°C and 60°C for even a short period of time (Azevedo et al., 2014).

In April and May of this year, the bacteria strain *E. coli* O157:H7 caused an outbreak in 26 states across the U. S. A few investigations have revealed that the contamination may have originated in the Yuma, Arizona, growing area. This outbreak, which started on April 10, is the biggest *E. coli* outbreak in the United States in a dozen years. In California, one person died. At least 14 of those affected experienced renal failure. The most prevalent *E. coli* symptoms are diarrhea, bloody diarrhea, abdominal cramps, nausea, and vomiting.

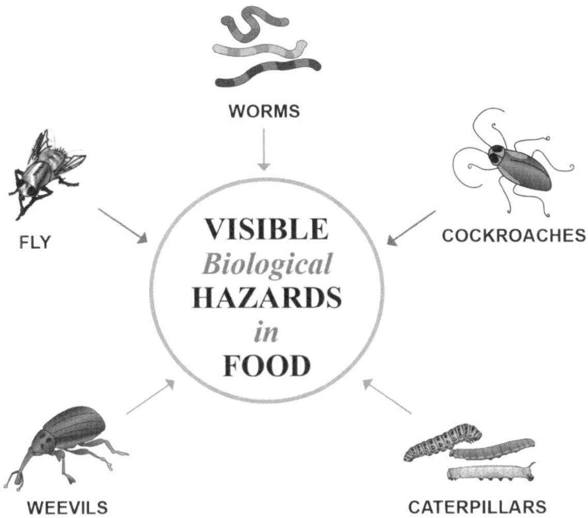

Figure 4.7: Visible biological hazards in foods.

Source: http://courseware.cutm.ac.in/wp-content/uploads/2020/06/session-4.pdf

4.5 FOOD SAFETY

Food safety refers to the guarantee that food is fit for human consumption in accordance with its destined use. The definition of two other concepts — toxicity and hazard — improves one's comprehension of food safety. Toxicity is a substance's ability to cause harm or injury under any circumstances. The relative likelihood of harm or injury occurring when a material is not used in the specified format and volume is referred to as the hazard. Physical, chemical, and biological hazards can all have a negative impact on consumer health.

A hazardous substance is any physical material that is not typically found in food that causes injury or illness and would include wood, stones, pest parts, hair, etc.

Chemical hazards are harmful compounds or chemicals that are knowingly or unknowingly added to foods. Pesticides, chemical contaminants, toxic materials, polychlorinated biphenyls, preservatives, food colors, and other additives are examples of hazards. Living organisms, including microbiological organisms, constitute biological hazards (Li et al., 2017). Food-borne pathogens are microorganisms that are connected with food and cause disease. Infections and poisoning are the 2 kinds of food-borne illnesses caused by microbial pathogens.

Food poisoning is caused by the ingestion of live pathogenic organisms, which grow in the body and cause illness. Salmonella is a perfect example. This organism can be found in the intestines of animals. Eggs and raw milk are also sources. Although heat kills Salmonella, insufficient cooking permits some organisms to survive. Cross-contamination is a common way for Salmonella to spread. This can occur when a cook cuts uncooked meat on a cutting board and then utilizes it for another food that does not require cooking, like salad. Salmonella can infect food if an infected food handler doesn't quite wash their hands with soap after using the restroom and then before touching food. Salmonella can multiply very quickly, doubling their count every 20 minutes. The symptoms of Salmonella infection include diarrhea, fever and abdominal cramps. Food intoxication: Some bacteria produce harmful toxins which are present in food even if the pathogen has been killed. Organisms produce toxins when the food has not been hot enough or cold enough. Toxins in food cannot be detected by smell, appearance or taste. Hence foods which smell and appear good are not necessarily safe. One example of such an organism is *Staphylococcus aureus*. Such organisms exist in air, dust, and water. They are also present in the nasal passage, throat and on skin, hair of 50 percent of healthy individuals. People who carry this organism, contaminate food if they touch these places on the body while food handling. Diarrhea is also one of the symptoms of this contamination.

Infestation can also be caused by parasites, such as tape worm infestation in pork. Furthermore, food can be riddled with pests and insects. Biological hazards are a major reason of food-borne illness among the exposures. Despite all efforts in the field of food safety, microbiological pathogens remain a serious major worry, and new pathogens spring up on a regular basis.

Figure 4.8: Infestation of foods.

Source: https://www.muhafizpest.com/food-which-will-cause-pest-infestation/

Human hosts, animal hosts and their interplay with humans, the bacterium itself, and the ecosystem, such as how food is grown, processed, treated, and preserved, are all significant considerations in the rise of pathogens. Changes in host vulnerability, for instance, caused by malnutrition, age, and other factors, can enable the development of new infections in susceptible populations. Genetic interaction or mutations in microbes can result in the formation of new strains with the ability to cause illness (Lam et al., 2013). Pathogens can arise in new populations or geographical regions as a result of changes in eating behaviors, climate, mass production, food processing, and enhanced globalization of the food supply. Norovirus, Rotavirus, and Hepatitis E are just a few examples, accounting for roughly 70% of cases. New pathogens will continue to proliferate, necessitating the techniques to isolate, control, and identify their existence in foods. It is essential to comprehend the definitions of contamination and adulteration in the field of food safety.

Food safety is critical for the following reasons:

- You cannot eat or drink if the food or drink is unsafe to consume. The most obvious example is safe water to drink. People would never drink water from such a reliable source. The same principle holds true for food.
- Every day, people all over the world become ill as a result of the food or drink they consume. Food poisoning can be caused by bacteria, viruses, and parasites present in foods.
- There is no direct way to taste or smell anything out of the ordinary.
- Food poisoning can cause serious health problems, including kidney damage and death.
- This risk is particularly high for infants, pregnant women, the elderly, and the immune-deficient, particularly HIV patients and cancer patients (King et al., 2017).
- Food hygiene and safety keep germs from multiplying and approaching dangerous levels in foods.
- Promotes healthy family living on a daily basis.
- Maintaining one's health and avoiding the additional costs of drugs and medical check-ups. This amounts to billions of dollars annually lost revenue as a result of employee downtime.
- Hand washing is responsible for 33% of all food poisoning cases. As a result, it is critical to practice good personal hygiene. Even though we are taught this from a young age, hand washing is still a major issue in the kitchen.

Food safety is a scientific concept that describes how to handle, prepare, and store food in order to avoid food-borne illness. This contains a count of routines that must be followed in order to avoid potentially serious health risks. Food safety measures ensure that the food will not endanger the consumer when prepared and/or consumed as directed (Jin et al., 2020). Food safety is critical for the reasons stated: You cannot eat or drink if the food or drink is unsafe to consume. The most obvious example is safe water to drink. We would never drink water from an untrustworthy source. The same principle holds true for food. Every day, people all over the world become ill as a result of the food or drink they consume.

Figure 4.9: Following four simple steps at home—Clean, Separate, Cook, and Chill—can help protect you and your loved ones from food poisoning.

Source: https://www.cdc.gov/foodsafety/keep-food-safe.html

Food poisoning can be caused by bacteria, viruses, and parasites present in foods. Because you can't see, taste, or smell anything out of the ordinary, there's no way to tell if food is tainted right away. Food poisoning can cause gastroenteritis and dehydration, as well as potentially more serious health problems like kidney failure and death. is particularly important for those who fall into the high-risk category: Small children or babies, pregnant women, the elderly, and the immune compromised, particularly those with HIV infections. Food hygiene and safety keep germs from incrementing in foods and infecting people. It ensures healthy family living on a daily basis. Keeping oneself healthy and avoiding the additional expense of purchasing medication and ups. This is especially true in business. Companies around the world lose billions of dollars each year as a result of employee downtime.

Hand washing is responsible for 33% of all food poisoning cases. As a result, it is critical to practice good personal hygiene. This is stuff we are taught from a young age, but hand washing is still a major issue in the kitchen (Jespersen & Huffman, 2014).

- To offer a concept and vision about the essence of the food product and how it will be produced, including all of the R other factors that will be assessed in order to obtain a safe end product.
- Recognizing crucial areas in food processing and preparation that are needed for safe food production
- Identifying food safety hazards

- Be familiar with the implementation of efforts to manage these dangers where they are substantial.
- More efficient use of resources
- Centralization of hazard management facilitates auditing and inspection.

4.6 FACTORS INFLUENCING SAFETY

Foodborne illness stories have become far more common around the world. Is food less safe than it once was, and if so, what factors contribute to this? Some risks to food safety have existed since antiquity, while others have emerged as a result of changing demographic characteristics and ways of life, production procedures, and even the evolution of microorganisms (Jacxsens et al., 2010).

4.6.1 Contamination of Food

Foodborne diseases, or diseases associated with food, are defined as illnesses that affect a large number of people and are costly to treat. Foodborne diseases are caused by consuming contaminated food products. Contamination of food at any phase, from production to parasites, chemical agents, and toxins causes foodborne diseases.

4.6.2 Demographics

Infants, small kids, the elderly, pregnant women, those taking specific medicines, and those with illnesses like acquired immunodeficiency have a greater risk of becoming sincerely ill. According to one study, 90 percent of deaths with diarrhea as the root cause were for the elderly or children. They are more susceptible to bacterial infections. The human immune system, which is not developed properly at birth, matures gradually during puberty and then starts to decline after about fifty years old.

Figure 4.10: Demographics also affect other business factors such as pricing, packaging, and service offers. When marketing campaigns are aligned with the demographics of customers, sales will likely increase.

Source: https://corporatefinanceinstitute.com/resources/knowledge/economics/demographics/

4.6.3 Consumer Lifestyles

With the fast pace of life, we often eat meals on the go and spend less time preparing food, opting instead for restaurants, processed food, or ready-made meals. This means that eventually you enjoy your food, it has likely been moved, cooked, allowed to cool, preserved, transported once more, heated up, and handled by a number of people. Each processing step presents new hazards, potentially allowing pathogens to survive and grow (Jackson, 2009). When you add in the improper handling of food that takes place after a customer buys something and brings it home, the risk of illness rises. Roughly 20% of foodborne diseases cases are caused by food prepared at home. Experts believe that this figure is much greater, but that most individuals do not disclose illnesses caused by home-cooked foods.

4.6.4 Food Production and Economics

In the past, foodborne outbreaks were linked to local events like weddings or public gatherings where a significant number of individuals ate the same food. Food is created in entirely different ways nowadays than it was even a few decades ago. Locally grown, produced, and distributed food was once the norm. Food production is now more consolidated and on a bigger scale than previously. Products made in large quantities in a single processing plant are transported all over the country, and sometimes even around the world. A handling error will be felt across the country rather than just locally. Even the way farmers raise their animals can make a significant contribution to a rise in food safety issues. A vast number of animals are crammed together, putting a strain and weakening immune systems. This overcrowding also aids in the transmission of infections from one animal to the next. In the past, a sick animal would be kept relatively isolated, and if it became ill, it would join the rest of the flock or herd. However, the group as a whole has a closer animal.

4.6.5. New and Evolving Pathogens

Scientists only discovered four foodborne pathogens fifty years ago, and that number is now on the list. Three of the four pathogens that the CDC considers the most significant in inducing foodborne illness were not even recognized twenty years ago—*Campylobacter jejuni cayetanensis* first emerged in 1979 and is still poorly comprehended. As analytical techniques improve, scientists are likely to discover new foodborne pathogens. Pathogens, like organisms, are always changing. Scientists are finding that some bacteria can survive in previously thought-to-be-safe environments. *E. coli*, for example, has been linked to meat products such as salami, apple cider, unpasteurized milk, and was originally thought to be the cause of breakouts in acidic foods like salami and apple cider. *Yersinia enterocolitica* refrigeration temperatures are now understood (Hellwig et al., 2019). The processing step of cooling, storing, transporting again, reheating, and being touched by numerous people introduces new hazards that may permit the survival and development of bacteria. Add to that the poor handling of food that takes place after a customer buys food and heads home, and the risk of illness rises. Approximately 20% of observed foodborne illness cases are caused by food that is home cooked. Experts believe that this figure is much greater, but that most individuals do not report illnesses caused by home-cooked foods.

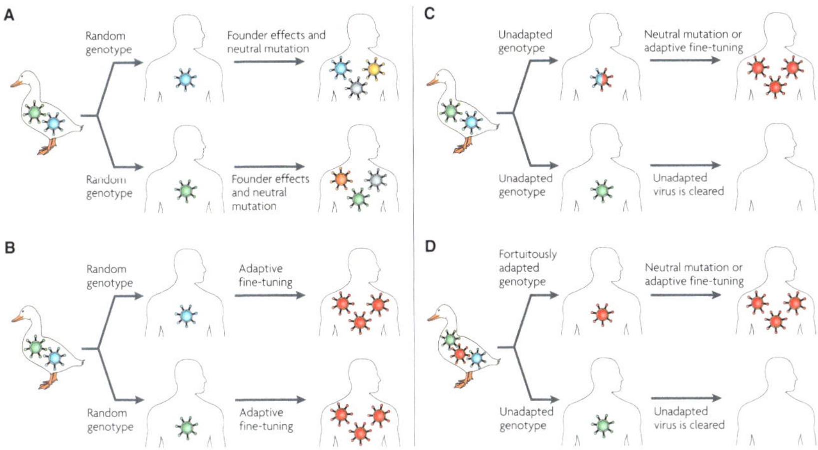

Figure 4.11: Different mechanisms of pathogen emergence in novel host species.

Source: https://www.embopress.org/doi/full/10.15252/embr.202051374

4.7 FACTORS THAT IMPACT FOOD SAFETY INSIDE THE 21ST CENTURY HOME

To comprehend the problems of food safety in the household, it is crucial to evaluate the pertinent elements that make up an average modern home in the early twenty-first century. It's also important to remember that, in so many parts of the globe, the home is a multipurpose configuration that hosts a range of events that can affect the need for and exercise of food safety (Antle, 2001).

The household is, first and primarily, a residence with inhabitants of various ages and wellbeing statuses. The count of immunosuppressed citizens is increasing in many parts of the world, and these folks are at a greater risk of acquiring a foodborne disease and having a much more aggressive disease outcome, along with other things. The number of immune deficient adults in the U.S. is approximated to be around 30 million. The older population is the largest rising portion of the population in many developed countries, such as the British Isles, where there are presently 10 million elderly people, the majority of whom live at home. To get a sense of the size of the immune deficient populations in the developing world, one only needs to study the consequences of the AIDS epidemic and the population with HIV/AIDS.

With a massive change in healthcare provision, the household is ever more serving as an augmentation of, or substitute for, conventional in-hospital care in the United States. 8 million patients are now being cared for at home, with 66 percent of them that are over the age of 65. This emphasizes the importance of food safety in the home once more, much as it would if these patients were also being cared for in a clinic.

The inclusion of young kids in home-based daycare is yet another growing home-based exercise that may have an impact on food safety in the home. In the U. S., 75% of children under the age of five are registered in daycare, accounting for 13 million preschool children and 6 million babies and children. The majority of that kind of child care delivered at home, with family and friends caring for 25% of all kids and in-home caretakers caring for 5% of all children (Hermann et al., 2018). There have been numerous reports of epidemics in children's day healthcare setting, including diarrhea, and the risk of infection spreading into the home such as through food is invariably enhanced in these cases. Furthermore, outbreaks in small and/or non - licensed residence daycare configurations are less likely to be disclosed and investigated, and as a result, recommendations and details about the importance of proper food safety processes may be missed.

The residence is regularly home to a variety of pets, varying from common to exotic, and so many zoonoses, many of which cause foodborne outbreaks, can be obtained from both. Salmonella and other enteropathogens have consistently been associated with domestic animals such as cats and dogs. Domestic cats and dogs can act as carriers for Campylobacter species, making them potential infection sources. Exotic pets can bring enteropathogens into the house with them. Cats and dogs are found in more than half of English-speaking households, with 14 million in the U.K., 60 million in the U. S., and an approximate 17.8 million domestic pets in Australia, with three out of every five households having at least one pet. The role of household pets in infants acquiring Salmonella species infections was discussed. In this study, it was discovered that infants were more likely to contract infections through close interaction with inanimate surfaces tainted by household pets, like floors, than through the ingestion of contaminated food.

Ultimately, when we assume food safety in the household, we generally think of meals that have been prepared and presented to the residents. Even so, we should keep in mind that the home kitchen can be used for food drives, churches and schools picnics, and other small home-based businesses that cook meals for catered events outside of the home (Baluka & Rumbeiha,

2016). Meals cooked at home are provided to a larger community in each of these instances. These catering operations are typically unregulated, occur in kitchens with insufficient equipment and facilities, and are staffed by people who have not completed food safety training.

In an investigation of home-based catering activities in thc U.K., it was discovered that 50 percent of the time, food was stored incorrectly in the home kitchen.

4.8 GLOBAL IMPACTS ON FOOD SAFETY IN THE HOME

Food safety in the home in the twenty-first century is also influenced by a number of global factors. Modernization of the food supply, in particular, has an impact on households all over the world.

Between 1983 and 2020, global meat demand is projected to double to 300 million metric tons, with the majority of the increase occurring in developing countries. Because meat processing may not always be well subject to regulation, home facilities may not be furnished for storing and preparing raw meats, and the workforce may not be acquainted with regular food safety regulations for meat handling and storage, the effect on food standards for homes in these countries could be substantial, particularly where large amounts of meat protein are consumed. The research on campylobacteriosis in developing countries mentioned earlier demonstrates the potential scale of the issue encompassing poultry production (Havinga &Verbruggen, 2017).

According to import statistical data, more than half of the fresh vegetables sold in developed markets come from emerging regions, provoking food safety specialists to joke that 'traveler's diarrhea' can be contracted by simply going to the local supermarket and returning home.

International travel & tourism to regions with poor food standards of cleanliness may have a negative impact at home. Yearly, 1.6 billion people travel by air around the world. Individuals who have suffered gastroenteric infections may bring these agents back into their residences due to rapid air travel, with the possibility for further propagation within the relatives and the inner circle, both directly through person-to-person interaction and obliquely through cross-contamination into home-cooked food. According to a WHO report published in 1995, 64 percent of tourists leaving Thailand in 1995 were experiencing diarrhea.

4.9 IMPROVING FOOD SAFETY IN THE 21ST CENTURY HOME

There are three main defenses against foodborne illness, according to legend. The first aims to improve the hygienic value of raw foods; the second utilizes food processing innovations like pasteurization and irradiation, as well as HACCP ideas; and the third focuses on food handler teaching, including that of home-based food handlers. There are a myriad of regional and international initiatives aimed at informing and educating the wider populace about the need to comprehend and practice food safety, such as the Canadian Partnership for Consumer Food Safety Education, the Partnership for Food Safety Education in the United States, the Food Standards Agency in the United Kingdom, and others. These initiatives are thorough and address legitimate fears, but they undoubtedly rely on web and print media, making them an inert form of data exchange that is likely to be adopted only by those people who are actively looking for information (Griesche & Baeumner, 2020). Food safety has conventionally been taught as a component of the home economics school curricula in many nations, but home economics has faded away from many national educational programs in recent years, and as a consequence, food safety is not instructed. Shifts in social structure, family meal procedures, and gender relations in the home and workplace have all contributed to a breakdown in the communication of data about safe dietary patterns within the group. With rising public concern about the prevalence of foodborne disease and the massive national financial impact affiliated with it in many states, mandatory food safety education initiatives in schools should be presented as a means of assertively educating and interacting with the citizenry on a basic health problem.

In practice, the aimed hygiene strategy provided and explained by the International Hygiene Forum in their Guidelines for prevention of infection and cross infection in the domestic environment is increasingly being incorporated into food safety knowledge and awareness. Focused hygiene is a risk-based strategy for home hygiene that is comparable to the HACCP approach used in the commercial food industry. The relative need for a hygiene intervention is determined by the origin of pathogens in the household, the potential paths of pathogen transmission within the home, and the probable risk posed by the transmission of these agents to a close relative. The danger of pathogen transmission to other sites and surfaces can then be reduced by recommending safe and efficacious hygiene practices eradicating pathogens from those substrates that pose a risk (Gill et al., 2007).

Now that you know what food safety is, let's talk about food quality.

Food Safety: The term "food quality" refers to the characteristics that influence a product's consumer value. This will include both negative characteristics like spoilage, contamination, adulteration, and food safety risks, as well as positive characteristics like color, flavor, and texture. As a result, it is a holistic concept that incorporates nutritional characteristics, sensorial properties (color, mouthfeel, shape, visual appeal, taste, flavor, and odor), social considerations, and safety. Safety is a prerequisite for quality and an attribute that precedes it. Various countries and public bodies have established food standards that food manufacturers are predicted to follow in order to guarantee that standards are met and of good quality (Fung et al., 2018).

As a result, all food service providers (those involved in pre-preparation, preparation/processing, packaging, and service) should follow good manufacturing practices to ensure food safety. The following are important considerations:

- Raw material and water quality
- Premises, employees, equipment, food storage and handling, and serving areas are all clean.
- Food storage at the proper temperature
- Food safety
- Excellent customer service.

4.10 FOOD SAFETY REGULATIONS

To integrate performance into every facet of food production and service, as well as to guarantee the source of hygienic, nutritious meals and to promote business between and within countries, proper food standards and control measures are needed. Four stages of standards are in place, all of which are well-coordinated.

- Company Standards: A company's own standards are created for its own use. They are usually copies of National Standards.
- National Standards: The Food Safety and Standards Authority, a national standards body, issues these

- Regional Standards: Legislation standardization bodies exist for regional groups with analogous geographical, climatic conditions, and other characteristics.
- International Standards: International standards are published by the International Organization for Standardisation (ISO) and the Codex Alimentarius Commission (CAC).

Food Standards, Quality, Research, and Trade: International Organizations and Agreements.

Since ancient times, governments around the world have attempted to establish and apply food standards in needed to shield consumers' health and avoid deceitful food sales practices. Several global organizations and agreements have contributed to improving food safety, quality, and security, as well as enabling research and trade. The following are some of the major organizations that are playing a vital role:

- The Commission on the Codex Alimentarius (CAC)
- The International Organization for Standardization (ISO).
- Organization for Economic Cooperation and Development

4.10.1 Codex Alimentarius Commission

CAC is an intergovernmental organization tasked with developing international standards for protecting consumer health and enable food and agricultural trade. In 2017, there were 187 member nations and one constituent organization (the European Community) in Codex. The Ministry of Health and Family Welfare of India is a member. CAC has established itself as the most important international bench mark for food standards developments. Codex Alimentarius, which means "Food Code," is an accumulation of worldwide assimilated Food Standards reported by the CAC (Fritsche, 2018). To safeguard consumers and guarantee fair practices in the food industry, the documentation contains Standards, Codes of Practice, Guidelines, and other guidelines. Codex Standards are used by various countries to enhance federal standards.

4.10.2 International Organization for Standardization (ISO)

The International Organization for Standardization (ISO) is a global, non-governmental federal state of national standards organizations (ISO member bodies). ISO's objective is to enhance the development agenda of uniformity and similar events in order to facilitate international trade in goods and services and to foster cooperation in the academic, scientific, innovation, and economic fields. ISO's work produces international agreements that are authored as International Standards. ISO 9000 is a global standard for quality requirements. It is involved with an organization's "Quality Management." The implementation of these standards is entirely optional (Ergönül, 2013).

4.10.3 World Trade Organization (WTO)

The World Trade Organization (WTO) was founded in 1995. The WTO's main goal is to facilitate smooth, free, fair, and predictable trade by prescribing trade agreements, resolving trade negotiations, and aiding countries with trade policy issues. Goods, services, and intellectual property are all covered by the WTO Agreement. A strong Food Control System is required to enforce the formulation and usage of standards. Food inspection and analytical capability are both necessary components of an effective meal control system (Bech-Larsen &Aschemann-Witzel, 2012).

Food inspection is used to ensure that products meet industry standards. All foods will be produced, handled, processed, stored, and dispersed in accordance with regulations and legislation. Food inspectors are appointed by government / municipal authorities to probe the condition of quality compliance to standards in their labs. Analytical capability: To conduct food analysis, well-equipped, state-of-the-art approved laboratories are required. In addition, well-trained personnel with knowledge of laboratory management concepts as well as physical, chemical, and microbial assessment of food, test foods, and food products are needed. To detect food residues, environmental chemicals, toxicants, microbial pathogens, food-borne viruses, and parasites, a wide range of advanced analytics is required.

4.11 FOOD SAFETY MANAGEMENT SYSTEMS

Food safety and quality concerns have evolved over time to include more than just the prevention of food-borne pathogens, chemical toxicants, and other hazards. A food hazard can enter/come into the food at any point along the food chain, so proper control is required throughout the chain.

A food hazard can enter/come into the food at any point along the food chain, so proper control is required throughout the chain.

- Good Manufacturing Practices (GMP)
- Good Handling Practices (GHP)
- Hazard Analysis Critical Control Points (HACCP)

Figure 4.12: US FDA scientist tests for Salmonella.

Source: https://en.wikipedia.org/wiki/Food_safety#/media/File:FDA_Lab_3000_(4494152579).jpg

GMPs are a type of quality control that ensures that manufacturers and processors take measures to make sure that their goods are safe. It allows for the reduction or elimination of contamination and false labeling, protecting consumers from being misled and assisting in the purchase of safe products. GMP is a useful business tool that helps manufacturers/producers improve their compliance and performance.

Good Handling Practices describe a thorough approach to identifying prospective sources of risk from the farm to the shop or consumer, as well as the methods taken to reduce the risk of contamination.

It ensures that everyone who handles food does so in a sanitary manner.

Hazard Analysis Critical Control Point (HACCP) stands for Hazard Analysis and Critical Control Points, and it is a method of ensuring food safety. HACCP is a method of food manufacturing and storage in which raw ingredients and each personal step in a process are scrutinized for their potential to support the growth of harmful microbes or other food hazards. It entails identifying hazards, assessing the likelihood of hazards occurring at each step/stage in the food chain — raw material sourcing, manufacturing, allocation, and food product usage — and identifying measures to mitigate hazards (EFSA et al., 2020).

Why should you use HACCP?

- It's a proactive strategy for ensuring food safety.
- End-of-product inspection and testing, while necessary, is time-consuming, costly, and only detects problems after they occur. HACCP, on the other hand, allows us to detect dangers at any point of the process or manufacturing in order to guarantee a high end product by taking necessary actions at the point of occurrence.
- It allows producers, processors, distributors, and exporters to make efficient and cost-effective use of resources to ensure food safety.
- The Food Safety Modernization Act of 2006 (FSSA) assigns primary responsibility for food safety to producers and suppliers via HACCP, GMP, and GHP. This is critical for the protection of consumers and international food trade.
- It ensures high-quality products on a consistent basis.

Food safety in the U. S. involves the processing, wrapping, and food storage in such a manner that food-borne disease is avoided within the country. Food safety laws in the United States began in the 20th century, when several outbreaks prompted this need for food litigation management in the food industry. Several federal agencies were established in the United States over the next few centuries in an exertion to better comprehend food pollutants and govern them. Since the turn of the century, many food-safety laws have been enacted and amended in the United States.

Following several deadly outbreaks in the early 2000s, the United States has lately given food safety a second thought. Incidents like the *E.coli* polluted leafy greens in 2006 draw focus to the food industry's regulation and food quality assurance. Many outbreaks have happened as a result of lax regulatory enforcement and a lack of decent testing of each batch of produced food. The majority of food-safety legislation was enacted in the aftermath of a deadly epidemic of a food-borne disease. Salmonella, *E.coli*, Listeria, Norovirus, Campylobacter, and *Clostridium perfringens* are the bacteria and viruses that account for the majority of food-borne illnesses. These can cause various diseases that have claimed the lives of many adults in the United States. In order to avoid and report accurately food-borne diseases in the U. S., food safety is required. In 2011, there were 9.4

million food-borne ailment instances in the U. S. Rather than serving as a preventative indicator against food-borne illnesses, pervasive food-borne epidemics usually prompt legislation (Diasolua et al., 2002).

Figure 4.13: Informing consumers of the proper ingredients in food helps avoid widespread food-borne illness.

Source: https://en.wikipedia.org/wiki/Food_safety_in_the_United_States

There were no laws concerning food, intentional additives, or unintentional pollutants added to food even before to 1900s. Upton Sinclair's novel The Jungle, authored in 1905, depicted the dreadful workplace conditions in the meatpacking industry. The public was outraged by his lengthy description of the poor quality of meat.

Following the reports of poor food quality, two acts were signed into law in 1906: the Pure Food and Drug Act and the Federal Meat Inspection Act. Food manufacturers were required by the Pure Food and Drug Act to only sell pure unbridled foods and to properly label them. The Meat Inspection Act led to the establishment of the Food Safety and Inspection Service of the United States Department of Agriculture, which oversees the meat production, poultry, and eggs, attempting to enforce regulated contaminants limits and quality of products (Cody et al., 2012). These laws have established a precedent in food regulation and have become a foundation for food safety in the United States. Amendments and changes have occurred since the signing of these two acts, but only within the framework established by the Pure Food and Drug Act and the Meat Inspection Act. These acts enabled the United States to rule in the case of United States vs. 95 Barrels Alleged Apple Cider Vinegar that the apple cider vinegar in discussion was misbranded because it was made from dried apples rather than fresh apples.

4.12 GOVERNMENT AGENCIES

The Food and Drug Administration (FDA), the Food Safety and Inspection Service (FSIS), the Center for Disease Control and Prevention (CDC), the State Department of Public Health, and the State Department of Agriculture are the three federal and two autonomous government entities in charge of food safety in the United States. These organizations focus on food production and distribution, ensuring that all food distributed to retailers, eateries, and buyers is safe and free of food-borne illness contamination. These institutions are the most active in governing food and avoiding food-borne illness in the United States, despite the fact that there are many other relatively small organizations that take part in the dispersion of safe food.

4.12.1 Food and Drug Administration

The Food Safety and Inspection Service (FSIS) monitors the 25% of food that the FDA does not regulate, which contains all egg, meat, and poultry products prepared and scattered in the United States. The Federal Meat Inspection Act of 1906, the Poultry Products Inspection Act of 1957, and the Egg Products Inspection Act of 1970, as well as several amendments to these acts, formed the foundation of FSIS. Every state in the US, such as Puerto Rico, imposes safe, wholesome, and properly arranged egg, poultry, and meat products. FSIS is currently a department of the United States Department of Agriculture.

In the United States, the FDA regulates over $2.5 trillion in consumer food, medical products, and tobacco as of 2018. Commissioner of the FDA is the position held by the FDA's chief. The Food and Drug Administration is currently under the direction of the United States Secretary of Health and Human Services (Chen et al., 2020).

The Bureau of Chemistry, a government division, was established in 1862. In 1927, the Bureau of Chemistry was divided into two divisions, one of which was known as the Food and Drug Administration. When it came to enacting the few regulatory food safety laws, the FDA had little direct authority as a new institution. Even so, in the late 1930s, an incident enhanced the FDA's food and drug control.

Sulfanilamide, a novel antimicrobial drug developed in the 1930s, became widely accepted. S.E. Massengill Co. in Bistol, Tennessee, had a chemist who wanted to find a way to liquify the non - soluble drug so that children and adults could take it in liquid form. The drug was dissolved in diethylene glycol (DEG), which is toxic. 1300 bottles of this "Elixir

Sulfanilamide" were scattered to consumers and clinicians without any toxicological testing. The drug was recalled as quickly as customers started dying after trying to take the elixir. The FDA conducted an investigation into the company and discovered no clinical trials or proof of more testing of the elixir (Bingham &Lavau, 2012).

The FDA did not have the authority to chastise the organization at the time. Charges were filed against the owner, and six months later, the Food, Drug, and Cosmetic Act was passed. This law required the FDA to certify all new foods, drugs, and cosmetics before they could be sold. This act gave the FDA enforcement and legal authority, which it has used to help monitor and control food and drugs ever since.

Figure 4.14: The FDA helps regulate contaminants in all types of food.

Source: https://en.wikipedia.org/wiki/Food_safety_in_the_United_States

4.12.2 The Centers for Disease Control and Prevention (CDC)

The US Public Health Service's Malaria Control in War Areas program was renamed the Communicable Disease Center in 1948. (CDC). On July 1, the Centers for Disease Control and Prevention (CDC) was founded in Atlanta, Georgia. One of the CDC's first goals was to completely eradicate malaria in the United States, which it actually achieved in 1951. When polio began to spread in the 1950s, the CDC began to develop tracking tools to monitor and record polio cases in order to aid in the fight against the paralytic and deadly virus's spread. In the 1970s, fewer than ten people in the United States had contracted polio. The CDC proceeded to endorse for other diseases and immediate public health issues after its name was changed to the Center for Disease Control (CDC) in the 1970s (Boudergue et al., 2009). When the Zika outbreak started in the United States in 2016, the Centers for Disease Control and Prevention (CDC) immediately activated their Emergency Operations Center and ended up taking proper measures. The CDC's main goal is to keep track of threats to Americans' health, safety, and security.

Figure 4.15: The Centers for Disease Control and Prevention (CDC) helps target public health issues including food-borne illnesses caused by contaminated food.

Source: https://en.wikipedia.org/wiki/Food_safety_in_the_United_States

CHAPTER 5

BEST PRESERVATION PRACTICES TOWARDS FOOD SAFETY

CONTENTS

Figure 5.1: Food preservation ensures food sustainability.

Source: https://www.britannica.com/topic/food-preservation

There are several practices used across the world to help maintain food safety. Food safety is defined as the routine practice in the preparation, handling and storage of food so as to prevent foodborne injury and illnesses. Currently, food preservation practices across the world have been modified to ensure food safety. Maintenance of food safety is very important at various levels in the community. All of these activities are geared toward making food safe for consumption. Food preservation practices have been made on the concept of food safety. The concept of food safety brings users to the knowledge that once food has been harvested, slaughtered or gathered, there are bacteria and enzymes that become active in food leading to the deterioration of food texture and composition rendering it unfit for consumption (Chandrapala et al., 2012).

The deterioration of food is known as decay and the final stage is food spoilage. Food is said to be safe for human consumption when it is free from substances such as micro-organisms, toxins and contaminants. When it is consumed the body develops undesirable reactions in the body. There are several benefits of food safety including effective utilization of food when consumed by the body leading to improved health and nutrition outcomes. It also leads to prevention of food-borne illnesses and it reduces food wastage.

Figure 5.2: Proper food preservation prevents food spoilage.

Source: https://link.springer.com/chapter/10.1007/978-3-030-24903-8_2

5.1 MINIMAL PROCESSING

Minimal processing is an optimized food preservation design in the aspect of nutrition, safety and sensory quality that does not make use of the addition of preservatives. This method of food preservation has gained popularity among consumers making it ascend to higher levels. This is because some of the activities under minimal preservation can be done at home. Over the years, users have an increased demand for fresh and natural-like foods in minimal processing technologies. Most of these technologies are known to delay the oxidation of products. There are several parameters and factors taken into consideration for food to be subjected to minimal processing processes. For instance, water activity is used in calculating the possibility of growth of microorganisms while the redox potential is used in finding the oxidation with regards to the gaining or losing of electrons (Carvalho, 2006). A major challenge in the food industry is maintaining both the quality and freshness of the product. There are different categories of minimal processing. The major categories include minimal processing to improve the organoleptic properties and maintain nutritional values with major concerns being microbiological safety, the use of mild heat treatment techniques such as post-harvest technologies and protective microbiological treatment and minimal processing to obtain synergistic effects which involve a combination of various techniques and methods. Both the protective microbiological treatment and post-harvest technologies are collectively known as novel preservation techniques.

Figure 5.3: Minimal processing does not use preservatives.

Source: https://foodyaari.com/minimal-processing-of-foods/

5.2 MINIMAL PROCESSING OF FRESH FOODS

Minimal processing is done to make fresh food available in the market by increasing food stability and nutritional properties. A variety of approaches have been used in preserving products on the basis of surface preservation by mild heat treatment, coating and packaging techniques while ensuring the nature of products is maintained. All of these methods have been useful in ensuring food safety. They are discussed as follows:

5.3 VACUUM PACKAGING

This method of preservation involves the sealing of products in an impermeable membrane after air has been removed fromthe membrane hence the name vacuum packaging while the process is known as atmospheric packaging. This method has been useful in increasing the stability of food product to levels more than that of foods that have been refrigerated. In the long run, it prevents the browning reaction that occurs on the surface of high perishable products during damaging or cutting. This method maintains food safety through the removal of oxygen. This is because oxygen is the major requirement for the growth molds, yeast and bacteria. As oxygen is removed through pumping, its levels drop causing an increase in carbon dioxide levels increase (Beulens et al., 2005). The presence of carbon dioxide does not create a suitable environment for the growth of spoilage microorganisms. For this reason, it is considered an economical method of improving the shelf life of different products. This method is used in the preservation of perishable foods. After this food has been packed through this method it is placed in refrigerators as a number of microorganisms are resistant to carbon dioxide. Vacuum pumping has several advantages including the fact that it reduces the shrinkage of products, it also reduces loss of moisture and the mass of the product does not affect until packages get damaged. Pouches used in vacuum packaging are made of the finest virgin material in a thickness suitable for handling, shipping and storing food with minimal care.

Figure 5.4: Vacuum packaging of meat.

Source: https://www.sciencedirect.com/topics/agricultural-and-biological-sciences/vacuum-packaging

5.4 GAS FLUSHING

This kind of packaging method is a modified form of atmospheric packaging that makes use of different inert gases and carbon dioxide and nitrogen. This process of food preservation is known as gas flushing. The gases used in this case play the role of an inhibitor for several spoilage microorganisms in several ways of elevating oxygen from the atmosphere of the product.

5.5 CARBON DIOXIDE ATMOSPHERIC PACKAGING

Carbon dioxide is used in food preservation as ithas an inhibiting action on pathogens. It is often used in preservation at certain concentrations. Usually, carbon dioxide is used in preservation in the range of 30-40%. When used in relatively large amounts, it leads to the discoloration of several products including meat. The carbon dioxide is usually absorbed by the food product resulting in pack collapse thereby affecting the moist rates. This problem is overcome by combining carbon monoxide within the limit of 35 ppm. A combination of carbon monoxide and myoglobin resultsin the production of carboxymyoglobin. This results in the production of a red pigment giving the color of bright cherry (Bolek, 2020). The discoloration of food is minimized by using this method. This is because this method will not affect several clostridia and it inhibits the growth of spoilage microorganisms.

Figure 5.5: Carbon dioxide does not allow the survival of microorganisms.

Source: https://www.google.com/amp/s/www.scotsman.com/business/consumer/what-is-co2-used-for-in-the-food-industry-and-is-there-a-carbon-dioxide-shortage-in-the-uk-3392412%3famp

5.6 NITROGEN PACKING

Nitrogen as a gas displays little or no antimicrobial activity on its own. The composition prevents the collapsing of packs as the composition cannot have a very low capability to dissolve in fat and water when subjected to atmospheric packaging. During this form of packaging, nitrogen is added by displacing oxygen in sufficient levels during packaging. It leads to the delay of oxidative rancidity and limits the growth of aerobic microorganisms. Nitrogen is mostly used for nuts and packing of perishable foods. It is useful in maintaining the flexibility of packaging and it also acts as a filler gas. Currently, there is no limitation in the usage of gas in several foods as there are no adverse effects on food. 100% of nitrogen is used when packing freshly cut vegetables and is then stored at low temperatures (Borchers et al., 2010).

5.7 ARGON PACKAGING

Several tests have been done on the use of argon for modified atmospheric packaging. In Europe, Argon is in packaging different kinds of foods and is used in preserving food for commercial purposes. There are a number of reasons why argon is used in preserving food. Argon is much denser than nitrogen. In nitrogen packaging, there are certain amounts of oxygen that remain and cause the oxidation of food. This is overcome by the use of argon and the spaces in between are completely filled. Argon is used in preventing the deterioration of food flavor through the inhibiting of microbial oxidase enzyme involved in oxidation.

5.8 CONTROLLED ATMOSPHERIC PACKAGING

Atmospheric packaging is used in maintaining the atmospheric composition. Controlling of atmospheric packaging entails the maintaining of different parameters including temperature and relative humidity. It also involves controlling the optimum range that is a major factor involving the instability of food in the area of quality and safety. A number of techniques have evolved over the years enabling the stability of packaging and food. It works by coating with several films and packing using gases.

5.9 EDIBLE FILMS

The nature of products can be maintained by coating the outer layer of food material is followed. Edible films are the materials used in enrobing materials. Food items are usually not consumed immediately after harvesting. There are several processes used in making edible films before it is consumed by consumers. These processes include transportation, storage, handling of products, and different stresses being applied on the products. The edible coating is used to prevent loss of product composition, nutrition and stability. The dibble coat slows down the deterioration. There are polysaccharide based films and coating. This process involves different polysaccharides coating food. Some of the formulations used include alginate, pectin, cellulose, chitosan and starch. The films provide both strength and structural integrity. The film also acts as a barrier for oxygen aroma and oils. However, there is a challenge of the hydrophobic nature that holds moisture. In such films, hydrogen is bonded network structure and low solubility (Bouzembrak et al., 2019). As it is tightly packed, oxygen acts as a barrier. There are protein-based films and coatings. Both vegetables and fruits easily go through oxidation. This is prevented by having the aroma nature and from the contact of oxygen proteins are used in coating. Materials used in coating include corn-Zein, casein, and whey protein and egg albumen. However, the coatings do not act as a moisture barrier. The moisture barrier is created by the presence of fatty acids. Whey protein is an example of a protein with fatty acids. Proteins gotten from milk are used in preventing the browning reaction in vegetables and fruits.

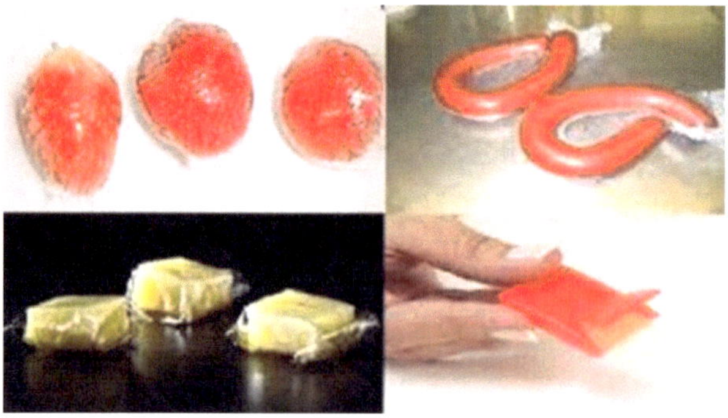

Figure 5.6: Edible films preserve foods for longer durations of time.

Source: https://www.researchgate.net/figure/Some-examples-of-edible-films-in-the-food-packaging-industry_fig2_338224773

There are lipid-based films and coating used in preserving food. There are different types of lipid-based materials including paraffin wax, selective types of oils and surfactants. When used in coating food, they act as effective barriers and protect foods from agents causing deterioration. This kind of coating is used in preventing the entry of moisture and water vapor as it has low polarity. The high solubility, microscopic pores and diffusivity are characteristics of partial oxygen barrier. This kind of coatings and films have several advantages including being opaque and inflexible. The final coating is the composite and is made up of various types of coating materials. Good compositions make the coating effective. There are different kinds of proteins that exhibit good mechanical properties that allow the development of a stable film system including plasticized protein. Research shows that protein casein from acetylated monoglyceride and milk from lipid for potatoes and apple that are slightly processed is stable for about 3 days from moisture loss and oxidative browning.

Minimally processed foods have quality attributes (Zanin et al., 2017). This is the quality of a product that serves an important point in the market of minimal processing foods. The quality attributes of the safety of the product includes color, flavor, nutritional value and appearance. All of these attributes are formed on the basis of the internal and external factors and is determined by the product's condition of pre and post-harvesting. The quality of food can greatly be evaluated by its appearance. Appearance is considered an important quality as it is useful in indicating the freshness of the product. Food can be classified according to the color dispersion, defects in shape, glossiness of exposure of the product and structure of the product. Changes in color occur when the product ripens while discoloration is as a result of enzymatic browning when vegetables or fruits are cut.

5.10 CANNING

Canning is a method of preservation that involves foods being placed in jars or containers and heating them at high temperatures that destroy micro-organisms causing food spoilage. During the heating process, air is driven out of the jar. After cooling, air is driven out creating a vacuum where it is sealed. The vacuum is useful in preventing the entry of air back into the jar. Canning is normally used in food preservation as it is suitable for low and high acidic foods, enhances shelf life and prevents contamination. There are a number of stages in the canning process. The first stage is cleaning and it involves raw food being passed through tanks of water or under high-pressure water sprays. The product is then cut, peeled, cored, sliced, graded, soaked and pureed (Xiang et al., 2019). When canning vegetables and fruits, blanching may be required by immersion in steam or hot water. This serves either as an additional or final cleansing step. Blanching causes the softening of vegetable tissues making them pliable enough for tight packing. This process also leads to the inactivation of enzymes which when present in food will cause undesirable changes in food before it is canned.

Figure 5.7: Canning involves food being placed in cans.

Source: https://www.google.com/amp/s/health.clevelandclinic.org/how-you-can-keep-at-home-canning-safe-avoid-botulism/amp/

The next step involves the filling of cans. This is often done automatically by machines. They may be a lineup of cans and the machine will fill in solid content. In most cases, there is a liquid accompanying the solid. By adding the liquid, a lot of air is replaced. Once the cans are filled, they are passed through a steam bath or hot-water bath in an exhaust box. Heating causes expansion in food and is useful in driving out the remaining air in cans. The cans are closed and sealed immediately after the exhaustion process. The sealing process is followed by sterilization. They are heated at very high temperatures for a given duration of time that ensures all the microorganisms are destroyed. The cans are then cooled in cold air or water. The final step is labeling (Wilcock et al., 2004).

There are different kinds of canning depending on the type of food. For instance, hot water canning is mostly used in the preservation of high acidic foods including pickled vegetables and fruits. Pressure canning is usually used in the preservation of low acidic foods including vegetables, dairy products, poultry, seafood and meats. The hot water canning method involves jars filled with food being submerged in boiling water at a temperature of about 212 Fahrenheit. It is then cooked for a given duration of time. Some of the tools used include a water-bath canning kettle. For the pressure canning method, jars filled with food are placed in 2-3 inches of water in a pressure cooker specially made for canning for at least 240 Fahrenheit. Such a temperature can only be attained when the pressure method is used.

Figure 5.8: Hot water canning is used in preserving highly acidic foods.

Source: https://ladyleeshome.com/how-to-can-tomato-sauce/

Canning methods are usually geared towards microbial control. Boiling, pasteurization, refrigeration, freezing, vacuum treatment and drying are some of the methods used to prevent food from spoiling. Food preservation methods often use antimicrobial agents. Sterilization is the only method can be considered perfect and dependable as a preservative. Under sealing of canned food, there are different types. The double seams though invented in 1888 have been modified and have the ability to provide an airtight seal to tin cans. By creating an air-tight environment, microorganisms are kept out of the can while food is maintained in the can and sealed inside it making the double seem can be known as the sanitary can. The can was traditionally cylindrical with a tin plate. The lids at both ends are attached using the double seam. The can is usually sealed impervious to contamination as two tight continuous folds are created between the can's cylindrical body and the two lids (Wertheim-Heck et al., 2019).

The method of double seaming makes use of rollers in shaping the can, lid and double seam. The sanitary can is made using a sheet of coated tin plate. The can body is created by cutting rectangles which are then curled around a die. It is welded together leading to the creation of a cylinder with a side seam. Rollers are used in flaring out one or both ends of the cylinder creating a quarter circle flange at the circumference. High levels of precision are needed during the process to ensure the perfect alignment of the welded sides. An occurrence of misalignment will compromise the integrity of the can. A die cutter is used in cutting a circle from the sheet. The circle is then shaped.

Canning is very useful in improving the shelf life of various foods. It is also safe for consumption as high levels of cleanliness are usually maintained during the canning process. If the user does not trust the level of cleanliness of industries, they can get fresh food products and pack them at home. Canned fruits and vegetables are known to have high levels of dietary fiber and vitamins. There are cases where canned food products have higher nutritional value compared to frozen or fresh food products. When foods are exposed to heating during the canning process, dietary fiber is said to be more soluble and can readily be fermented in the colon into gases and physiologically active byproducts. When compared to fresh tomatoes, canned tomatoes tend to have a higher available lycopene content. Usually, canned meat and vegetables are usually stocked up for emergencies. However, with canning high levels of keenness has to be maintained as any mistake could lead to contaminants finding their way into food. There may be cases where the can is not well vacuumed.

5.11 DRYING

Drying is a method of food preservation that has been used throughout history with sun drying being the most common and cheapest means of drying food. The use of drying as a method of food preservation dates back to 12000 B.C. This method of preservation involves food being dried, dehydrated or desiccated. Drying is usually effective in preserving food as it inhibits the growth of various microorganisms including mold, yeast and bacteria through the elimination of water. Water in food was traditionally eliminated through evaporation by using various drying methods including wind drying, smoking, sun drying and air drying (Wahyuni et al., 2019). Currently, there are more modernized forms of food drying which include electric food dehydrators and freeze-drying. They are able to speed up the drying of food while ensuring consistent results. There are different kinds of food that are preserved through drying. A good example is meat which has played a significant role historically more so in Europe. The European diet greatly depended on dried cod also known as bacalhau or salt cod when salt was applied to fish and stocked in their absence. It was considered the main source of protein and was a major economic force within the triangular trade.

Figure 5.9: There are different types of drying.

Source: https://www.google.com/amp/s/food.ndtv.com/food-drinks/all-about-sun-drying-meaning-benefits-and-how-it-works-2239586/amp/1

Dried fruits were also consumed in the course of history as they had high levels of sugar and a sweet taste. When dried, they had a longer shelf life. There are different ways of using dried fruits. For instance, dried plums become prunes and grapes become resin. Dates and figs after drying are transformed into different products and can be eaten as they are, rehydrated or used as ingredients for other recipes. Vegetables are also dried through freeze drying. They are often used in the military, by hunters and backpackers. Onions and garlic can also be dried. Other foods that are dried include some groups of fungi and mushrooms (Tirado et al., 2010).

There are different methods used in drying food. For instance, electrical dehydrators are used when home drying meat, fruits and vegetables. Other than electrical dehydrators, foods are dried through wind or sun-drying. When drying certain foods, some preservatives can be used. They include BHT, BHA and potassium metabisulfite. In the absence of preservatives, the dried food products may need refrigeration or freezing to increase their shelf life. Industrial drying of food is often achieved through freeze-drying. This involves food being flash frozen. They are then placed into a reduced-pressure system. This leads to water being sublimated directly from the solid into the gaseous phase. Freeze-drying is usually more expensive when compared to traditional dehydration techniques. It also leads to the mitigation of the changes in nutritional value, texture and flavor. The other widely used industrial method of drying food is convective hot air drying. Usually, the industrial hot air dryers have a simple and easy design, they can easily be designed, construct and maintain. It is also very affordable and has the ability to retain a large amount of food's nutritional properties when dried using appropriate drying conditions.

Food can also be dehydrated through irradiation. Irradiation is effective in food preservation and uses x-rays, ultraviolet light and ionization radiations with the ability to penetrate to the point of sterilization. This method is mostly used in preserving food for astronauts and people who are at a high risk of microbial infections. Food can also be dried through hurdle technology. This technology is a combination of a number of food preservation techniques (Todt et al., 2009). The technology makes use of a low dose of multiple food preservation techniques to ensure food safety and it has a good appearance and texture. Packaging of dry food is useful in ensuring effective food preservatives. There are various ways of packing dehydrated foods including vacuum sealed, inert gases or gases useful in regulating respiration, growth of microorganisms and biological organisms.

Figure 5.10: A conventional oven can be used in drying food.

Source: https://www.whirlpool.com/kitchen/cooking/wall-ovens/single-oven/p.5.0-cu.-ft.-smart-single-wall-oven-with-true-convection-cooking.wos72ec0hs.html?bvstate=pg:2/ct:r

When drying food at home, one can use the conventional oven. For those who are new to food drying. However, using the conventional oven is much slower compared to the electric dehydrator. However, the individual will have to place little or no investment. Certain food can be dried through the conventional oven. They include meats, seafood, fruit skins and low moisture foods including berries, bread cubes, potatoes, herbs and meaty tomatoes. There are a few steps followed when drying food are as follows:

Trays used in drying foods are first prepared. It is followed by the preparation of foods that are to be dried. There is a variation in the preparation methods depending on the type of food. For vegetables and fruits, preparation mostly involves food being washedand are then halved or quartered and in other cases, they are sliced. Lightly-colored fruits and vegetables are usually steam-blanched to deactivate the enzymes and also prevent browning in lightly colored foods. They are then patted dry. For fish or meat, they are often marinated and in other cases, they are dried before cooking. The food is then preheated by using a gas or electric oven which is usually set to the lowest temperature setting. The oven temperatures are then usually maintained between 125 and 145 Fahrenheit (Sinha, 2009). An accurate thermometer is used in checking the temperature of the oven. Temperature is then decreased by propping open the oven door with a folded towel or wooden spoon. However, this method is not safe forsmall children. Drying is sped up through the maximization of air circulation. This is done by placing a fan on a chair near the propped-oven door. This blows away the hot escaping air. Airflow is promoted by opening nearby windows and doors.

Food is then dried till it is pliable or crisp. Food is usually dried depending on the preference of the user. This means that there can be a fluctuation in drying lengths ranging from a few hours to 24 hours. There are a number of factors that influence food drying time including the type of food, the size of portions to be dried, the method of drying to be used and the weather. Drying is also done when the atmosphere is less humid. There is also a manner in which food is dried. Electrical food dehydrators are usually more preferred forfood drying. This is because they are less dependent on weather and is known to produce consistent results. This makes electrical appliances easier than other methods. A good electrical drying appliance has good temperature control and good air circulation. When dried foods are to be used as snacks, they are drieduntil they are pliable. The same case applies to foods that are to be used for longer durations of time. For some foods such as tomatoes, once they are dried, they are grounded to make powders. Lower shelf lives are noted in less-dry products. When foods are completely dried, they are able to last for several months.

After drying food, they have to be well conditioned and stored. When foods are dried, they will have different moisture rates. When there are significant amounts of water in food and it is not well managed then it will lead to the development of mold and contaminate the whole batch. This requires measures to be put in place which include conditioning of food before storing. The conditioning process leads to the equalization of moisture such that excess moisture is transferred to drier places and finally, moisture is distributed throughout the batch. Dried foods are conditioned by placing them in tightly closed containers at room temperature. The contents of the container should be shaken every day for a week. This is done by opening the container and after shaking, it is closed. During the conditioning process, if moisture is on the inside of the container then the food isn't properly dried. To ensure food safety, it is important to ensure that containers used in storing dried foods are clean.

After the conditioning process, food is stored. This involves food being placed in airtight containers so as to enable food to last for longer durations of time. Usually, food is packed in small portions which reduce the number of times the package is reopened. Air entering the containers can be reduced by taping over it or using a desiccant to absorb oxygen. Stored food is usually kept at constant temperatures.

5.12 FREEZE DRYING

Freeze drying is a method of food preservation that involves freezing of food, lowering pressure and removing ice through sublimation. It is very different from dehydration by most conventional methods that evaporate water using heat. This method makes use of lyophilization to lower temperatures. Water is extracted from food through the use of a pressure vacuum. The collected vapor then collects on a condenser and turns back into ice and is then removed. Freeze drying was first applied after the Second World War. It was used in preserving food without the need for refrigeration. Among the first food products to be dried was coffee which paved the way for fruits, vegetables, meats, dairy products and herbs to be dried. Freeze-dried foods were very popular among the military and astronauts. Initially, the freeze dried foods were difficult to rehydrate (Sadiq & Beauchemin, 2017). This led to the modification in the process in which freeze-dried meals are heated. There are four main stages of freeze drying foods namely pretreatment, freezing, primary drying and secondary drying. The pretreatment process involves any method used in treating foods before they are frozen. Pretreatment may include concentrating food products, formulation revision, increasing the surface area or decreasing the high-vapor-pressure solvent. Pieces of food are often IQF treated. This is useful in making food pieces free flowing before freeze drying. For food to be pretreated, knowledge of freeze drying is used. There are a number of factors such as product quality that are also considered.

Figure 5.11: Freeze drying involves food being dried under low pressure.

Source: https://www.saintytec.com/food-freeze-dryer/

Pre-treating is followed by freezing and annealing which involves food being cooled down to temperatures below the triple point which is the lowest temperature in which solid, liquid and gas phases coexist. This ensures that sublimation rather than melting occurs. Freeze drying is done faster and effectively when large ice crystals are used instead. The large ice crystals are useful in forming a network within the product. This facilitates the fast removal of water vapor during sublimation. By freezing a product slowly, large crystals are produced. Annealing involves the products being frozen slowly or being cycled up and down in temperature. The most critical phase is freezing as the method can impact the speed of reconstitution, product stability, duration of freezing-drying cycle and appropriate crystallization. There are two types of drying namely primary and secondary drying (Smith & Pillai, 2004).

Primary drying is the phase that involves pressure being lowered and enough heat is supplied to the material causing ice to sublimate. The sublimating molecules' latent heat of sublimation is used to calculate the amount for sublimation. During this kind of drying, about 95% of water in the material is sublimated. This phase is usually slow as heat is to be supplied in heat in small amounts to prevent the material's structure from being altered. Pressure is also controlled by the application of a partial vacuum. The vacuum speeds up the sublimation process. This makes it useful as a deliberate drying process. The secondary drying processes are done with the objective of removing unfrozen water molecules in food as ice was removed in the primary drying phase. Secondary drying is usually governed by the adsorption isotherms. Secondary drying involves temperature being raised higher than in the primary drying phase. This is done so as to break the physico-chemical interactions between water molecules and frozen material. In this stage, pressure is usually lowered to encourage desorption.

Some foods preserved through freeze drying include coffee. By nature, coffee contains aroma and flavor qualities caused by the Maillard reaction. Robusta coffee beans that have been freeze-dried tend to have high levels of amino acids such as phenylalanine, lysine and leucine. Some of the essential amino acids that contribute significantly to taste are preserved. In fruits, conventional dehydration causes berries to have lower quality. This is attributed to their delicate structure and the high moisture levels. Freeze-drying of strawberries is known to increase their quality and other qualities such as color, flavor and ability to re-hydrate are maintained. There are different types of freeze-dryers that can be used in drying food. Some of the essential components in freeze dryers include vacuum chambers, process condensers, shelves, shelf fluid systems, control systems and vacuum systems (Sargeant et al., 2009).

Some of the advantages of freeze-drying are that it is a better choice of food dehydration as it is able to preserve food quality. This means the various attributes of food such as bioactivity, re-hydration and aroma. It helps increase the shelf life of food brought about by low processing temperatures in conjunction with the rapid transition of water by sublimation. The processing conditions ensure the minimization of deterioration reactions such as protein denaturation, enzymatic browning and non-enzymic browning. Freeze-drying increases the shelf-life of food products to more than 12 months. Food products after drying are not easily or fully-rehydrated. When re-hydrated, it has poor quality. Products that have been freeze-dried are porous in nature allowing them to be fully re-hydrated. This is useful in making ready-to-eat instant meals. Freeze-drying also ensures that there are minimal deterioration reactions allowing food to maintain its nutrient levels and its color. It also enables food to maintain its original texture and shape.

5.13 FREEZING

Freezing is considered the easiest and most convenient method of food preservation. Compared to other methods of food preservation, freezing is the least time-consuming method of food preservation. Most foods are known to freeze well in the exception of those that have a high water content, cooked starchy foods and cream-based items. When fresh, food stuffs contain enzymes which are compounds that help with the ripening and maturing of food products. When frozen, the enzymes in food products slow down the enzyme action but it does not stop it. When the enzymes are not inactivated, they lead to changes in color and flavor in food. It could also lead to a loss of nutrients. Freezing is also known to stop microorganisms. However, it does not destroy microorganisms. The cases of food spoilage and occurrences are reduced. When storing vegetables, it is important for the enzymes to be inactivated.

Figure 5.12: The refrigerator is an equipment used in freezing food.

Source: https://www.lowes.com/pd/Samsung-19-6-cu-ft-French-Door-Refrigerator-with-Ice-Maker-Fingerprint-Resistant-Stainless-Steel-ENERGY-STAR/1003081538

Blanching is very essential in obtaining top quality of frozen vegetables. The blanching process is usually done before the vegetables are placed in freezers. This is useful in ensuring that enzymes are inactivated. Blanching involves vegetables being exposed to boiling water or steam for a given duration of time. This is followed by rapid cooling in cold water that is below 60 Fahrenheit for the same duration of time to ensure that food is not cooked. Blanching is also useful in destroying microorganisms on the surface of vegetables. Once blanched, some vegetables such as spinach and broccoli become more compact. Depending on the type of vegetable, there is a recommended number of times for blanching vegetables. The recommendations should be followed to prevent over-blanching that may cause the product to be cooked and also causes a reduction in food flavor, color and nutrients. On the other side, under-blanching causes the stimulation of enzyme activity and leaves food at a poor stage than food that has not been blanched (Redmond & Griffith, 2004).

Recently, individuals have adopted the use of microwaves in blanching. This, however, is not recommended as it is known to produce uneven results brought about by the varied heat patterns within a microwave oven. When using microwaves using blanching, one should work with small quantities at a time. This means that no time will be saved when large quantities of vegetables will have to be blanched. Like vegetables, fruits also have enzymes that can cause browning and loss of vitamin C. However, they cannot be blanched like vegetables they are instead exposed to ascorbic acid that controls the enzymes in frozen fruits. This can be done at home as there is commercially available ascorbic acid that can be accessed by home owners. There are cases where lemon juice and citric acid is used in preventing the darkening of fruits. However, they may not be as effective as ascorbic acid. Browning can also be controlled by packing fruit in sugar or sugar syrup. This method is also not as effective as using ascorbic acid.

By preserving food through the use of freezing, changes may be noted in products including the development of rancid flavors. These flavors are noted when fats in products such as meat are exposed to air over a given duration of town. These flavors can be prevented through the utilization of a storage method that limits air access to foods. For this reason, it is required that excess air is removed from freezer bags or containers so as to reduce air contact with food. For foods to be stored for longer durations, there are certain temperatures that they are stored in. For instance, both fruits and vegetables should be maintained below zero degrees. A freezer thermometer is often used in checking whether the temperature is appropriate.

When food is stored at temperatures above 0 degrees Fahrenheit, then they are more likely to deteriorate and will demonstrate shorter shelf lives. It is also important to ensure that foods are stored in a constant temperature as fluctuations in freeze temperature can cause ice in foods in the freezer to thaw and when temperatures drop they freeze. Such an occurrence causes the smaller ice crystals to become larger hence cells in food are damaged and the product becomes mushier. When the surfaces of the product experience moisture loss and ice crystals evaporate from it then a freezer burn is created characterized by a gray-brownish spot on tissues becoming dry and tough. When food has been freeze burned, it is likely to develop off-flavors but when consumed, it will not cause illness. Freezer burn is prevented by packing food in heavyweight materials (Rendueles et al., 2011).

When storing foods in freezers, they should be properly packed to protect their nutritive value, moisture content, color and flavor. For these reasons, there are certain characteristics used in defining the most suitable packing material which include the level of resistance to vapor and moisture, leak-proof, resistant to soil and grease, is not susceptible to becoming brittle and cracking at low temperatures, has the ability to protect foods from absorbing other odors and flavors, easy to seal and easy to label. Before food is packed for freezing, it is important to ensure that they are all cooled. Food is usually packed in quantities that are usable for a single meal. Cold foods should be tightly packed into a container. When packing, it is important to ensure that there is some head space for food as it will expand on freezing. The amount of space left varies depending on the size of the container and food. When food is packed in bags, it should be pressed out to remove the excess air before the bag is sealed. Food packages are usually labeled and dates placed on each pack. Food is easily frozen by spreading the packages among already frozen food.

When there is a power outage for more than 24 hours or the freezer is broken, it is recommended that one uses dry ice in moving food from one region to another.

5.14 FERMENTING

Fermentation of food has been done throughout history in preserving food. It is used today in the preservation and production of food such as wine, cheese, and yogurt among others. Foods that have been fermented are known to be rich in beneficial probiotics and are known to have numerous health benefits. Fermentation of food during preservation mostly involves the conversion of carbohydrates to carbon dioxide and alcohols through the use of bacteria, yeasts or a combination. The use of bacteria and yeast leads to the creation of anaerobic conditions. For this reason, the fermentation process is said to be very simple. It can easily be defined as the chemical conversion of sugars into ethanol. The science of food fermentation is known as Zymology. In the historical study of fermentation, the catalysts used in the process were known as ferments and were found within yeast cells. They were known to function within living organisms. Under fermentation, the action of microorganisms is said to be desirable. It is usually employed in the leavening of bread made using yeast. It is used in sour foods during preservation as it is known to produce lactic acid.

Figure 5.13: Brine is usually used in the fermentation process.

Source: https://www.fermentedfoodlab.com/ways-to-keep-veggies-under-the-brine/

There are five main reasons why fermentation is carried out. It helps enrich food Diets through the development of a diversity of textures, aromas and flavors. It also enables the preservation of substantial amounts of food through alkaline fermentations, acetic acid and essential amino acids. It facilitates the biological enhancement of food substances with vitamins, essential fatty acids, essential amino acids and lactic acids. It facilitates the elimination of anti-nutrients and decreases the amount taken for foods to be cooked (Qiao et al., 2018).

Fermentation of food is known to bring out safe food products. This is because, during fermentation, lactic acid bacteria consume carbohydrates or sugars in food resulting in the production of flavor compounds and acids. Among other food preservation methods, fermenting is among those that can be done at home. While fermenting food, it is important to ensure that a safe and effective environment has been created. This is useful in facilitating the growth of desirable lactic acid bacteria. The fermentation process needs the presence of sufficient water which can be generated or added from vegetables after salting. There should also be sufficient nutrients that are usually provided by the vegetables, absence of air or oxygen and suitable amounts of salt. Salt concentration tends to vary depending on the recipe of food. During the fermentation process, there is the production of lactic acid and beneficial bacteria which compete for and destroy pathogens that could lead to foodborne illnesses when consumed. It is important to note that good food handling procedures should be followed to the latter. This is to ensure that food is not contaminated.

When the ingredients used in making food are contaminated by dirty working surfaces or dirty hands then the food sample will undergo the fermentation process but will not be safe for eating. For instance, when fermenting food such as sauerkraut, one starts by choosing the freshest cabbage. If the cabbage is old by some days, then water should be added to ensure that sufficient amounts of water are generated after salting. The cabbage is then pounded or squeezed, the type of salt used during the fermentation process is very important. During this fermentation process, iodized salt is not used. Pure salt is often advised. The best salt to be used in the fermentation process is the picking salt. Salt is often measured by weight rather than volume. There is a given ratio of the amount of salt to be added depending on the amount of food (Piggott & Marsh, 2004).

When fermenting cabbage, the container used should be large enough to accommodate the batch being fermented. A fermenting crock is often used. Before it is used, it is cleaned using hot soapy water and later rinsed with hot water. Fermenting weights are also useful as they hold down cabbage during the brine process. The brine used in the process is usually made up of 3 quarts of water with 4.5 tablespoons of pickling salt. Usually, the cabbage is cleaned, it is shredded, slated, pounded and finally liquid is added into the container until it covers the cabbage while the weights hold the cabbage down. To prevent the entry of insects, a towel is placed over the container. The fermentation container is placed at a temperature of 68-75 Fahrenheit. When exposed to warmer temperatures, food spoilage may occur or the fermentation process will take place for longer durations of time.

When the fermentation process starts, there are three main stages that will follow in the presence of a desirable temperature. The first stage begins with *Leuconostoc mesenteroides* initiating the fermentation process. It leads to the production of carbon dioxide as oxygen in the jar is replaced. The next step involves *Lactobacillus plantarum* and *Lactobacillus cucumeris* to continue the fermentation process which takes about 10-30 days. The last stage involves *Lactobacillus brevis* finishing the fermentation process within less than a week. There are different types of food that can be preserved through the fermentation process. For instance, with beans, there is the soy sauce, natto, miso, fermented bean curd, kinama and the iru. Under grain, some of the foods that can be preserved include chicha, boza, bread, beer, sake, ogi, rice wine and murri. With vegetables, there is the Indian pickle, mixed pickle, kimchi and tursu. Mead and metheglin are some honey types preserved through fermenting.

5.15 PICKLING

Pickling is a food preservation method that extends food shelf life through anaerobic fermentation in brine or food being immersed in vinegar. This kind of food preservation usually affects the flavor and texture of food. Food that has been pickled is then referred to as a pickle or pickled food. Eggs, dairy, fish, meat, fruits and vegetables are some of the pickled foods. Food is usually placed in a solution of a pH of 4.5 or lower. Such pH usually does not support the survival of most bacteria. Pickling as a method of food preservation allows the preservation of food for months (Pressman et al., 2020). When preserving food through brine, antimicrobial spices and herbs such as cloves, cinnamon, garlic and mustard seeds may be added. The brining process tends to be simple if the food contains sufficient amounts of moisture. In such a case, brining is achieved through the addition of dry salt. This is well illustrated by Korean kimchi and sauerkraut which are brined through the addition of salt on vegetables which will draw out excess water. Natural fermentation then takes place at room temperature. It usually involves lactic acid bacteria providing the required acidity that will ferment food.

In other cases, brining is achieved by placing the vegetables in vinegar. Similar to the canning process, the pickling process usually includes fermentation. However, food should not necessarily be sterile before it is sealed. The salinity or acidity of the solution, level of oxygen deficiency and the temperature of fermentation will determine the rate at which microorganisms dominate. It, therefore, determines the flavor of the end product. When both temperature and salt concentration is low, the solution is dominated by *Leuconostoc mesenteroides*. This results in the production of a mix of aroma compounds, alcohol and acids. On the other hand, presence of high temperatures and high salt concentration *Lactobacillus plantarum* dominates resulting in the production of lactic acid. In the production of most pickles, it starts with Leuconostoc and eventually changes to Lactobacillus with increased acidity. The process of pickling food is as follows:

Vegetables and fruits are submerged in brine with the amount of salt varying depending on the type of food. It could also be shredded and salted which is the case for the sauerkraut preparation. Flat stone layers are used in holding the vegetables underwater. Other methods involve the use of a tight lid or a lid with ananairtrap. Lids with the ability to release pressure may cause a buildup of carbon dioxide. There are cases where kahm yeast or mold is formed on the surface of vegetables. Although kahm is harmless, it is known to cause an off-taste. It can easily be removed with little effect on the pickling process. Chemical pickling involves vegetables or fruits being placed in sterile jars along with vinegar, brine or with diluted vinegar. The type of process used in priming food is often selected depending on the type of vegetable and fruit. For instance, placing of foods to be pickled in sterilized jars filled with water or vinegar is often used with foods that have high natural water content (Pollard et al., 2016).

Commercial pickling involves the use of preservatives such as EDTA and sodium benzoate. These preservatives are useful in enhancing shelf-life. Fermentation pickling usually involves food producing its own preservation agent. This kind of pickling involves Lactobacillus bacteria being the preservative agent as it produces lactic acid. To promote a crisp texture in food, Alum is often used. The United States Food and Drug Administration has approved its use as a preservative. When pickles are not fermented, they are known as refrigerator pickles. Most of the refrigerator pickles are made by marinating vegetables or fruit that has been seasoned in a solution of vinegar. For such foods to be stored for longer durations of time they are usually subjected to a canning process or stored under refrigeration. The Japanese Tsukemono is known to use a variety of pickling ingredients depending on the type and pickles are produced by the combination of the ingredients with vegetables to be preserved and the mixture is then placed under pressure.

5.16 CURING

Curing is a method of food preservation that forms part of the flavoring process of vegetables, fish and meat. This method of preservation is done through the addition of salt. It is usually done with the aim of drawing moisture out of foods through the process of osmosis. Curing of food is known to increase the solute concentration thereby decreasing its water potential. This renders food inhospitable for microbe growth known to cause food spoilage. Curing as a method of food preservation has been used in preserving food during the late-19th century. It was the primary method of preserving both fish and meat. The earliest form of food curing is dehydration. Usually, the curing process usually involves smoking, spicing, cooking and the addition or a combination of nitrite, nitrate and sugar.

Preservation of meat generally is composed of a set of treatment processes useful in preserving the properties of meat such as the texture, taste, properties and color of partially cooked, raw or cooked meats while ensuring that they are edible and safe for consumption. This method of food preservation is a dominant method of preserving meat for several years (Pappalardo et al., 2017). However, the development of other methods of preservation such as synthetic preservatives and refrigeration has been used to supplement it. Initially curing of meat was done to prevent diseases and to deal with issues of food security. Currently, curing of meat is usually done to increase the cultural value and desired outcomes of taste and texture of food. In less developed countries, curing is mostly done in meat production, transport and its availability.

CHAPTER 6

FOOD SAFETY REGULATIONS & STANDARDS

CONTENTS

Food poisoning has been a health risk since the dawn of civilization, and many of the food safety challenges we face today are not novel. Despite global efforts to enhance food safety, foodborne illness remains a serious public health concern in both developed and developing nations. Every year, 1.8 million people die from diarrheal illnesses, the majority of which are caused by eating or drinking contaminated food or water. The majority of foodborne diseases may be avoided by properly preparing meals.

The World Health Organization has long recognized the need for food workers to be trained on food safety obligations (WHO). WHO published the Ten Golden Rules for Safe Food Preparation in the early 1990s, which were extensively translated and replicated. However, it quickly became clear that a more basic and all-encompassing plan was required (Nesbitt et al., 2014). For more than a year, food safety experts and risk communicators collaborated to create the Five Keys to Safer Food poster, which was released in 2001. The Five Keys to Safer Food Poster condenses the Ten Golden Rules for Safe Food Preparation while also giving extra context for the suggested actions listed in the guidelines.

Figure 6.1: A food worker.

Source: https://shrm-res.cloudinary.com/image/upload/c_crop,h_705,w_1254,x_0,y_50/w_auto:100,w_1200,q_35,f_auto/v1/Legal%20and%20Compliance/supermarket_worker_wearing_mask_jx8fe7.jpg

6.1 A POSTER EXHIBITION ON THE FIVE KEYS TO A HEALTHIER FOOD SUPPLY

The Five Keys to Safer Food emphasize the need for hygiene, separating raw and cooked foods, accurate cooking, maintaining safe food temperatures, and using safe water and raw materials. The picture has been translated into more than a dozen languages to promote the WHO's message about food safety.

Based on established scientific concepts, modern food safety laws were designed to safeguard customers from foodborne sickness and death. In today's information age, with 24-hour news cycles and social media, expectations have shifted dramatically from a century ago. Consumers are usually made aware of possible food safety hazards before the firms involved are.

A lack of adequate food management may pose a significant health risk to the general population. Imports from foreign nations must be regularly examined to ensure that they fulfill food safety requirements and have not been tampered with (Meng & Doyle, 2002).

Governments throughout the world are revising and updating their laws. Food safety legislation becomes more outcome-based and transparent when industry-leading best practices are embraced. Another endeavor to ensure that commodities travel freely across the supply chain is the harmonization of food safety rules across national borders.

Every country has its definition of what constitutes a threat to its population. Both the United States and Canada have made significant modifications to their food safety rules as a result of the Food Safety Modernization Act and the Safe Food for Canadians Act. Countries ranging from Germany to New Zealand to China place a high value on food safety and the enforcement of their respective national legislation in this area.

The information below refers to current food safety regulations and the authorities in charge of enforcing them. Food Safety Modernization Act of the Magnificent United States of America: The Food Safety Modernization Act (FSMA) was passed by the FDA in the United States to regulate food safety. It grants the FDA broad authority to investigate private-sector food safety issues.

Figure 6.2: FDA manages food safety through the modernization act.

Source: http://www.qualityassurancemag.com/FileUploads/image/FDA_FSMA_act.jpg

Canada

According to the Safe Food for Canadians Act, Canadians may be confident that their food is safe (SFCA). The Canadian Food Inspection Agency issued the SFCA (CFIA). It strengthens food safety regulations, equips inspectors and labs with the resources they need to execute their jobs more efficiently, and educates consumers on the significance of food safety.

Europe

The European Food Safety Authority (EFSA), Europe's food safety authority, is headquartered in Brussels. The European Food Safety Authority is Europe's risk assessment agency for food and feed safety (EFSA). Experts at the European Food Safety Authority (EFSA) work collaboratively with national agencies and communicate freely with their stakeholders to give independent scientific advice and honest information on current or future hazards. The European Food Safety Authority (EFSA) is an independent non-governmental organization funded by the European Union's budget that is responsible for verifying the safety of European food (Mensah & Julien, 2011).

Australia and New Zealand are separated by a single border.

Food safety regulations Australia Within the country of New Zealand, there is a country named New Zealand. This bi-national government institution develops and administers the Australia New Zealand Food Standards Code. Food additives, food safety, labeling, and genetically engineered foods are all addressed in the Code. In Australia and New Zealand, state and territory governments, as well as food regulatory agencies, apply and interpret the Code.

Argentina

SENASA (Argentine National Food Safety and Quality Control Service/ ServicioNacional de Sanidad y CalidadAgroalimentaria) is a self-governing government institution in Argentina that monitors, regulates, and certifies animal and plant-derived goods. It is the organization's responsibility to safeguard the public from sickness and to eradicate it. Agriculture, Livestock, Fishing, and Food (SENASA) is a legal entity within the Ministry of Economy's Agriculture, Livestock, Fishing, and Food Secretariat.

Brazil

Several organizations and institutions work together to ensure the safety of Brazil's food supply. The major duty for monitoring agricultural commodities rests with the two National Sanitary Surveillance Entities (MAPA and MS), as well as other government agencies (ANVISA). MAPA's rules cover, among other things, animal goods, fresh fruits and vegetables, alcoholic beverages, juice, grain, cereals, seeds, and animal feed (including pet food). ANVISA is primarily in charge of enforcing processed food rules. MAPA and ANVISA regulations are available online. The search engine SISLEGIS on the MAPA website provides up-to-date rules on MAPA-regulated items (McLaughlin et al., 1999).

China

The China Food and Drug Administration (CFDA) arose from the old State Food and Drug Administration (SFDA). China's Food and Drug Administration was renamed and reformed in March 2013, elevating it to the status of a ministerial regulatory body. The Food and Drug Administration of Canada (CFDA) combined multiple overlapping entities into a single agency to streamline food and drug safety regulation. The State Council of the People's Republic of China, which is the competent authority for drug control in mainland China and is responsible for the general administration of food, health food, and cosmetics safety, reports directly to the CFDA.

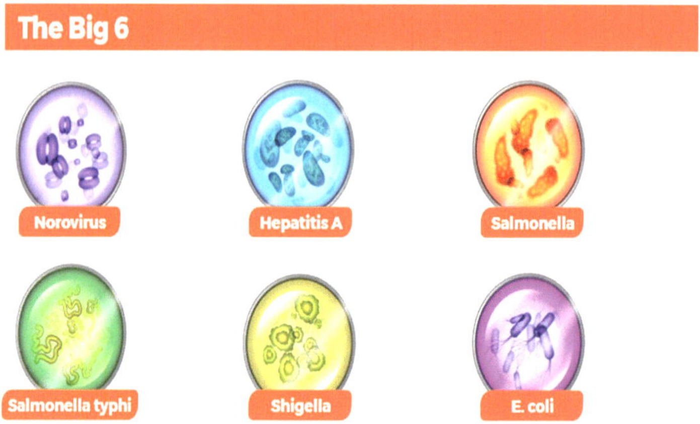

Figure 6.3: The Big 6 Foodborne Illnesses.

Source: https://alwaysfoodsafe.com/images/uploaded/87368307_3546747458699995_6928252925324034048_n.png

Foodborne illnesses are ubiquitous around the world and have a large human cost in terms of mortality and misery. According to the World Health Organization, poisoned food causes 1.5 billion episodes of diarrhea in children each year, resulting in more than three million early deaths (WHO) (WHO). Both affluent and poor nations are afflicted by these deaths and illnesses. Foodborne diseases, according to the Centers for Disease Control and Prevention (CDC), cause 76 million illnesses, 325,000 hospitalizations, and 5,000 deaths every year among the country's 290 million residents. Over one million children under the age of five die each year in Southeast Asia as a result of diarrheal diseases caused by contaminated food and water. Adulteration of food with hazardous compounds, whether accidental or planned, can result in serious public health consequences. • Wheat seeds poisoned with methyl mercury were unintentionally given to rural Iraq during the winter of 1971-1972. A total of 50,000 individuals were exposed to the contaminated bread, resulting in 6,530 hospitalizations and 459 fatalities. In Spain, poisoned rapeseed oil killed approximately 2,000 persons and damaged another 20,000, many of whom experienced severe paralysis. In 2002, when rat poison was purposefully used to infect bread in China, approximately 200 children fell unwell and died (Ambrus, 2016).

6.2 PERSONNEL-RELATED EXPENSES

Numerous nations lack effective monitoring and reporting methods for recognizing and tracking foodborne diseases. As a result, statistics on foodborne diseases are incredibly restricted, and modifications are necessary to better the identification of these illnesses' causes.

Foodborne illness symptoms can range from trivial to life-threatening. While nausea and diarrhea are the most typically reported adverse effects, others include kidney and liver failure, brain and neurological illnesses, and even death. For example, infection with *Listeria monocytogenes* has a mortality rate of 20 percent to 30 percent in the elderly and pregnant women. Foodborne illnesses can have long-term repercussions such as reactive arthritis and paralysis. Infants and small children, pregnant women, the immunocompromised, and the elderly are particularly vulnerable to foodborne sickness, which can be lethal. Foodborne illnesses are a significant cause of malnutrition in developing areas, impacting newborn and child growth as well as disease resistance.

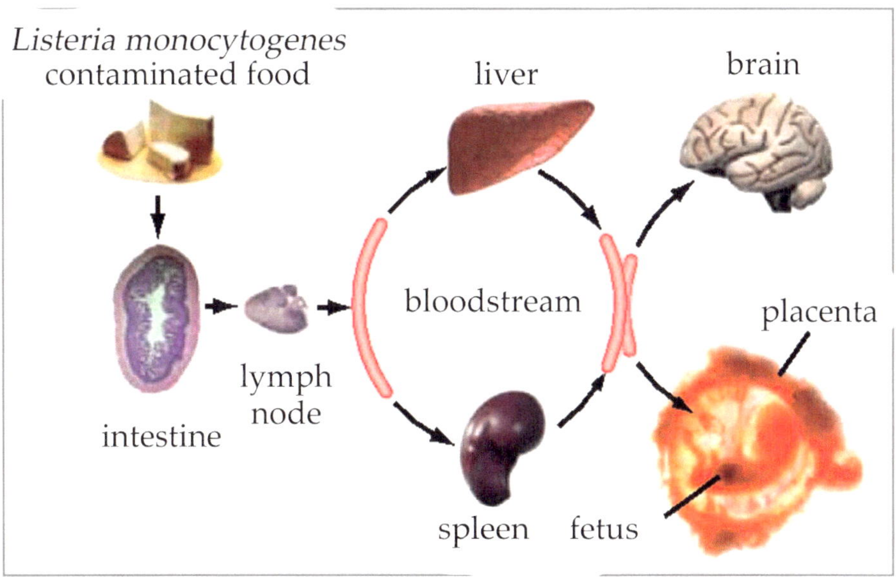

Figure 6.4: Infection with *Listeria monocytogenes.*

Source: https://www.pnas.org/cms/10.1073/pnas.1112371108/asset/afd-ecd90-301c-4357-95a5-b8efb8fd4d34/assets/graphic/pnas.1112371108fig01.jpeg

Malnourished newborns and children are more vulnerable to a range of ailments, such as respiratory infections, which can lead to increased malnutrition and disease. Each year, between 12 million and 13 million children die as a result of malnutrition and illness. Those who survive may experience a stop in their physical and mental growth, limiting them the capacity to fulfill their full potential in society. Production expenditures Foodborne illnesses have a severe impact on the economy. Consumer expenditures include medical, legal, and other bills, as well as time away from employment and education (MacLaurin, 2001). For many consumers who live on a subsistence level, a loss of revenue due to foodborne sickness might continue the cycle of poverty. Chronic illnesses generated by contaminated food, such as reactive arthritis or temporary paralysis, can be much more harmful than the original ailment and contribute greatly to medical costs and missed work. National governments suffer the expenses of greater medical expenditures, epidemic investigations, food recalls, and a loss of customer faith in the goods. Foodborne diseases impose a further burden on already overworked and underfunded healthcare systems in underdeveloped nations. The most accurate estimates of the economic cost of foodborne disease come from wealthier countries:

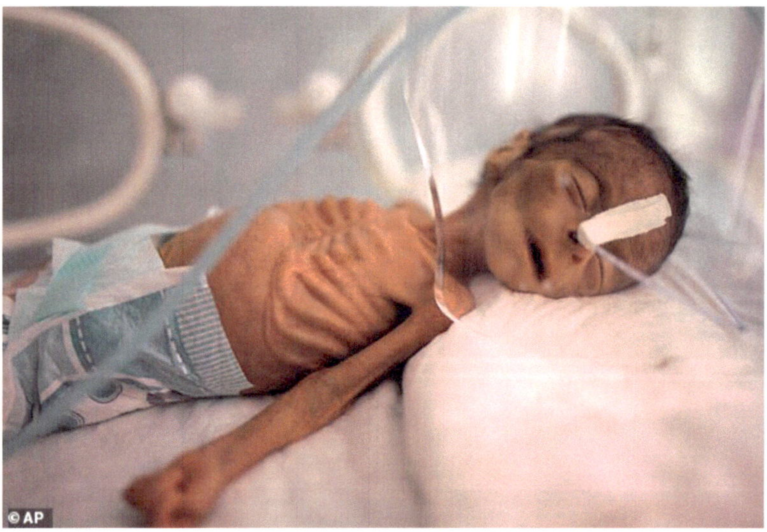

Figure 6.5: Malnourished newborns.

Source: https://pbs.twimg.com/media/EKQaHlaXYAAXiB1.jpg

In the United States, a government study of seven foodborne illnesses anticipated a cost of \$5.6 billion to \$9.4 billion in lost work and medical costs. The annual costs to the European Union's healthcare system as a result of Salmonella infections are expected to exceed EUR 3 billion. In Australia, the yearly cost of an estimated 11,500 daily cases of food poisoning was estimated to be AU \$2.6 billion. The cost of caring for and treating patients with the new Creutzfeldt-Jakob disease (vCJD) in the United Kingdom is expected to be more than £45,000 per case from the time of diagnosis, with an extra £220,000 reimbursed to each family under the government's no-fault compensation program (Mou & Lin, 2014).

Countries endure economic expenditures as a result of the globalization of food commerce when unsafe food results in lost exports. For example, a cholera epidemic in Peru in 1991, caused by the bacteria *Vibrio cholera* polluting water and seafood, lost the country more than \$700 million in missed fish and fish product exports. Due to a Cyclospora (a protozoan illness) pandemic in Guatemalan raspberries in 1996 and 1997, the number of Guatemalan raspberry growers had declined considerably from 85 in 1996 to three in 2002. Finally, the detection of bovine spongiform encephalopathy (BSE) in their cow herds resulted in losses of \$5 billion for the Canadian beef industry and \$2.6 billion for the US cattle business in 2004 (Lytton, 2017). Tourism is a major source of money for many countries. Being classified as a "traveller's diarrhoea" refuge might impact the country's reputation as a holiday destination and have severe economic implications.

6.3 POLITICAL CONSEQUENCES

Concerns about food safety can have serious political consequences. BSE has destroyed more political and structural ruin in Western Europe than any other food or agricultural problem. When BSE was identified in Germany in early 2001, the agricultural and health ministers both resigned, and the agriculture ministry was reformed to be more consumer-oriented. Food regulation activities in the United Kingdom were moved from the Ministry of Agriculture, Fisheries, and Food to a new, independent food authority, the Food Standards Agency. Similar state entities have been developed in several parts of Europe to enforce adequate food safety laws and restore public trust, and a European Food Safety Authority has been established (Li et al., 2017).

Current and future problems with food safety concerns differ by place owing to variances in economic levels, habits, local circumstances, and government infrastructures. The link between the food producer and the client is generally intimate in developing nations. There are less prepared and packaged meals; most fresh food is exchanged in traditional markets, and street merchants offer most of the food consumed outside the home. Perishable meals are often made and enjoyed swiftly, whereas prepared dishes are rarely stored.

Food safety concerns in these countries include improper agricultural chemical use, untreated or partially treated wastewater, and the application of sewage and animal manure to crops, a lack of food inspection, including meat inspection, inadequate refrigeration, and a lack of hygiene, including a lack of clean water supplies. The engagement of a nation in the global food market, as well as investment in agricultural capital, grow in parallel with its economic development. As a result, buyers may purchase both popular and unusual items at any time of year.

Figure 6.6: Animal husbandry.

Source: https://upload.wikimedia.org/wikipedia/commons/thumb/0/0b/Cattle_Feedlot_near_Rocky_Ford%2C_CO_IMG_5651-2.jpg/300px-Cattle_Feedlot_near_Rocky_Ford%2C_CO_IMG_5651-2.jpg

According to the WHO, the following trends may exacerbate food safety problems in both developed and developing countries:

Animal husbandry has evolved and altered over time. To increase output, modern intensive animal husbandry methods were used. As a result, Salmonella and Campylobacter have developed and become more widespread in flocks or herds of all main producing animals (poultry, cattle, and pigs). Increased antibiotic use has been linked to an increase in antibiotic-resistant bacteria as a result of overcrowding in so-called "factory farms." As a result of BSE, various research on the best methods to feed animals has been undertaken (Azevedo et al., 2014).

Agronomic process changes: Toxicological risks associated with fresh fruits and vegetables are increasing as a result of agricultural practices such as the use of manure, chemical fertilizers, untreated sewage, and pathogen-infested irrigation water; Fruit and vegetable-related epidemics have been on the rise in several locations, particularly as transportation and availability of imported fruits and vegetables has improved, providing customers with the more fresh product throughout the year. Several recent Cyclospora infections in North America and Canada connected to raspberries and lettuce, as well as over 9,000 cases of an *E. coli* O157:H7 epidemic in Japan in 1996, are examples. Because of increased international trade, germs may traverse borders more readily. If food is left out for extended periods, infection and time/temperature damagecan increase the risk of foodborne disease.

Figure 6.7: Plant Pathogens.

Source: https://d32ogoqmya1dw8.cloudfront.net/images/integrate/teaching_materials/food_supply/student_materials/figure_8.2.22_pumpkin_plant_744.png

As a result of greater trade, new and undiscovered foodborne dangers may come into touch with people who haven't evolved tolerance to such pathogens. Technological innovations in agriculture and food processing allow for the introduction of new foods to the market, as well as the introduction of new dangers owing to advancements in preservation, packaging, shipping, and storage. The increased use of refrigeration to lengthen the shelf life of ready-to-eat meals, for example, has aided the spread of *Listeria monocytogenes*. Many individuals in their communities are concerned about irradiation and genetically modified (GE) plants and animals. People are living longer lives and surviving chronic medical illnesses that were formerly fatal as a result of medical advancements. This makes impoverished populations more vulnerable. By 2025, about one billion people over the age of 60 will be living in emerging countries, with the developing world accounting for two-thirds of them. In certain countries, one out of every four individuals is now at risk of contracting a foodborne disease. When a person becomes infected with a foodborne disease in one country, they can spread it hundreds of kilometers to another, making the illness more hazardous (Lam et al., 2013).

Consumer expectations and lifestyle changes have a large influence on the occurrence and kind of foodborne infections. Seasonal dinners are popular among customers all year. Food produced outside the home makes for a bigger proportion of food spending in many developed countries. With increasing urbanization in developing countries, street food has become an integral component of many people's daily diets, particularly in the developing world. As a result, outbreaks linked to food produced outside the home are becoming more widespread in a variety of contexts. Bioterrorism: In recent years, there has been an upsurge in suspicions of food adulteration by terrorists, criminals, or other antisocial organizations, necessitating additional preparation processes. The WHO states that "the key to preventing food terrorism" is to "enhance existing food safety systems." Food law, food monitoring and surveillance, food inspection, foodborne disease surveillance, and education and training must all be appropriate and up to date to develop national food safety systems. The report contains contextual information. In the next sections of the article, food safety concerns in developed and developing countries are addressed, as well as the reactions of governments, international organizations, and consumer groups. Even though filthy water is a major cause of gastrointestinal illness in many parts of the globe, this study focuses on foodborne hazards (King et al., 2017). Furthermore, it is not the intention of this inquiry to cover every possible place. Its material is based on a range of regional and international articles. Certain criteria were used by the World Health Organization to assign states to regions (WHO). Consumer organizations in each region assessed and commented on each regional report, and their recommendations and critiques were included.

6.4 WESTERN PACIFIC REGION

The Western Pacific Region is riddled with food safety risks. Some Western Pacific countries have excellent food safety procedures in place, while others lack clear regulations. Many governments are unable to strictly enforce food laws duc to a lack of resources. Australia and New Zealand have more established systems due to their affluence and reliance on agricultural exports. Food safety was acknowledged as an important public health concern during the WHO Regional Committee's 52nd session in 2001, and a regional plan was developed. Infections caused by tainted food in the western Pacific Ocean In 2002, a rival poisoned bread with rat poison, sickening over 200 students and killing 38 of them (Jin et al., 2020).

Figure 6.8: Western Pacific Region.

Source: https://www.researchgate.net/profile/Amado-Iii-Tandoc/publication/256404088/figure/fig1/AS:613864924254222@1523368211763/Countries-and-areas-in-WHOs-Western-Pacific-Region.png

This was one of numerous disastrous epidemics that occurred in the Western Pacific Region. A tragic disaster occurred in Tangshan, a Nanjing neighborhood. The Snow Brand Company's Osaka facility was the cause of a food poisoning incident that killed 14,780 people in 2000, making it one of Japan's deadliest food poisoning outbreaks. In 1996, severe *E. coli* infections from white radish sprouts infected 9,578 people in Japan, the majority of who were students. In 1988, a Hepatitis A epidemic linked to poor eating infected 292,000 individuals and killed nine. According to a study, there is an upsurge in foodborne disorders caused by biological hazards in the Western Pacific region. Campylobacter, enterohemorrhagic*Escherichia coli*, Cyclospora, and Listeria have all increased. Recent incidences of sickness have been connected to novel *Vibrio cholerae* (O139) strains as well as drug-resistant Salmonella bacteria (Jespersen & Huffman, 2014).

Food sickness induced by chemicals is relatively uncommon. Pesticide residues, for example, pose a significant health danger in Vietnam. A poison known as Ciguatera is one of the most prevalent dangers to fish in the Pacific. Many people in the Western Pacific Region have been infected with botulism, often after consuming fermented foods or sausages. Food safety is a problem that both producers and consumers commonly misunderstand. Between 1996 and 1998, for example, 70 people died as a result of drinking pesticide-tainted rice wine. Consumers in the Western Pacific area face a health risk from live animals and poultry. In 1998, 105 individuals died as a result of a Nipah virus outbreak in Malaysia. Pigs and pork are commonly related to the Nipah virus (Jacxsens et al., 2010). A new strain of avian flu, known colloquially as "bird flu," has been detected in the Western Pacific area. People who came into contact with contaminated birds or cared for sick family members who were infected died or became ill as a result of avian flu. These epidemics had a significant impact on public health, as well as the cattle and meat product industries. In addition, in February 2005, Japan uncovered the first human case of mad cow disease in the region.

6.5 WESTERN PACIFIC REGION

Concerns about food safety in the Western Pacific

A variety of variables, according to WHO research, contribute to foodborne disease in the Western Pacific Region. Intensive agricultural methods include the use of animal waste and slaughter by-products as feed, the abuse of antibiotics and other toxic chemicals, and the frequent use of growth hormones. It is unsafe to collect fish in contaminated waterways or to employ illegal poisons and explosives in aquaculture. Contaminated items are widely spread across the industry when contamination occurs in major processing plants with extensive distribution networks. In an attempt to increase the shelf life of food, new packaging and processing technologies are being exploited.

Figure 6.9: Foodborne diseases kill approximately 50,000 people annually in Western Pacific Region.

Source: https://www.theguardian.com/environment/2020/jan/23/appetite-for-warm-meat-drives-risk-of-disease-in-hong-kong-and-china

Some Pacific Island communities have not appropriately addressed the hazards of marine toxin contamination linked with the capturing of particular fish species. As a result of rising urbanization, the availability of ready-to-eat and fast food has expanded, resulting in a scarcity of waste disposal facilities, safe drinking water, and sanitary services. Women have fewer educational opportunities and less decision-making authority in the house when it comes to food preparation. Cross-border travel of tourists, migrants, and employees has increased, as has global commerce in food and feed. Food labeling regulations are lax for both domestically produced and imported goods (Jackson, 2009).

6.5.1 Western Pacific Regional Policies and Action Plans

Food safety regulations are non-existent in some western Pacific countries. It is fairly unusual to find public health campaigns and plans of action that have nothing to do with food safety. In most circumstances, while establishing food safety measures, government policy does not have a specific purpose in mind. Regularly, several government entities deal with food safety issues. Some nations have a complex network of food safety rules and regulations in place, whereas others do not. Because inspection actions in the Western Pacific Region are not coordinated, the industry is exposed to either numerous inspections or no inspections at all. When there aren't enough inspectors, it's difficult to adequately coordinate national operations. Even though there are numerous food safety inspectors, their training is sometimes poor, and there is little evidence available on how inspection impacts food safety. The New Zealand Food Safety Authority (NZFSA) began a performance-based verification program for a variety of processed foods in 1999. The frequency and severity of inspections are determined by the food safety risk of the product and the performance of the producer (Hellwig et al., 2019).

Many countries have robust export certification processes in place. As a result, export products must fulfill higher criteria than domestic items. As a result, the Western Pacific Region's national capacity to conduct food laboratory tests varies. A lot of countries lack the capabilities needed to detect and isolate foodborne diseases and chemical risks. In Australia, for example, foodborne disease investigations have changed dramatically: more complex and extensive investigations are becoming the norm; outbreaks are becoming more widespread; smaller outbreaks are being identified, and more outbreaks involving contaminated imported products are being identified. Many governments are likewise decentralizing their laboratory systems, resulting in a multiplicity of overworked and underfunded state, provincial, and municipal laboratories. Only a few nations in the Western Pacific Region engage in monitoring programs, and they typically cover only a limited number of pollutants. The most typically collected data include pesticides, heavy metals, and mycotoxins. There are not always controls in place to assure the quality of the lab's work. In many nations where foodborne disease reporting is required by law, monitoring mechanisms are insufficient.

People who become ill as a result of food poisoning are not required to report it. Food safety specialists in just a few countries may employ formal training programs based on process control concepts (such as Hazard Analysis and Critical Control Point systems) (Hazard Analysis and Critical Control Point systems). Only a few Western Pacific countries support food safety initiatives and a standard implementation. Even though governments frequently target consumers for education and training, consumer organizations are rarely recognized as reliable information sources or change agents in food safety concerns. ASEAN established a regional food safety communication and coordination network (Hermann et al., 2018). Australia and New Zealand developed the binational FSANZ to set food standards for composition, labeling, and pollutants, including microbiological limits, that apply to all commodities manufactured or imported into Australia and New Zealand.

6.5.2 Consumer Advocacy Groups in the Western Pacific Area

Consumer organizations in the Western Pacific have been very active in educating the public about food safety issues. They strive to educate the public about food-related problems and push for food labeling rules that will allow consumers to make more informed decisions through consumer campaigns, publications, and newsletters. Consumers in the Western Pacific Region also conduct product testing to assure the safety and cleanliness of a variety of commodities in the region. Throughout these testing, antibiotic residues were discovered in fresh fish and shellfish, as well as dyes, heavy metals, and chemicals (Havinga & Verbruggen, 2017).

6.6 SOUTH-EAST ASIAN REGION

Although South-East Asia achieved tremendous nutritional gains in the 1990s, food safety did not. The people of South East Asia are concerned about the safety of their food supply. Population growth, particularly in metropolitan areas, has exacerbated food safety issues in many nations. The food safety concerns here have been addressed through the formation of a regional plan.

Figure 6.10: South-east Asian region.

Source: https://www.iapb.org/wp-content/uploads/2020/09/iapb-regions-south-east-asia.png

6.7 FOODBORNE INFECTIONS IN SOUTHEAST ASIA

Most Southeast Asian nations are afflicted by illnesses carried by tainted food. Bacterial contamination of food and water is one of the leading causes of diarrheal fatalities and diseases. Every year, children under the age of five die in this region from diarrheal infections caused by contaminated food and water. However, cholera has been a serious public health concern in this part of the world for decades, notably in India and Bangladesh. Cholera outbreaks are frequently triggered by polluted water, but street vendors and restaurants may also be selling infectious food. Food safety is a problem in Southeast Asia. Southeast Asia's health difficulties are aggravated by a lack of safe drinking water, increased use of pesticides and other chemicals in agriculture, and a lack of consumer and producer education. Rapid urban population expansion has resulted in extreme poverty, overcrowding, and inadequate sanitation in many nations. As a result, there has been an increase in food safety concerns (Griesche & Baeumner, 2020).

In virtually all of these nations, street food sellers and food service firms are a growing and significant element of the food supply chain. If proper precautions are not taken to assure the safety of foods throughout manufacturing, storage, distribution, and presentation, a substantial source of foodborne disease may ensue. Many nations, like India and Nepal, have health and food safety regulations prohibiting street food sellers from operating, however, these regulations go mostly unenforced due to a shortage of workers. Most Southeast Asian countries have few laboratories capable of diagnosing major foodborne dangers, and when they do exist, the high cost of testing is an impediment. In nations with a formal framework for food regulation, enforcement is frequently weak due to a lack of infrastructure and personnel.

Policies and plans for Asia are being developed and implemented across the area.

Southeast Asian countries agreed on a ten-point strategy to reduce foodborne infections in 1998. It's also worth mentioning that the strategy's objectives have only been met on an irregular basis. Even though every government has a large department dedicated to dealing with food safety issues, only a few nations have implemented food safety regulations (typically the Ministry of Health). Food-borne illness monitoring is also unreliable. Southeast Asian countries have begun to collect statistics on disease and mortality rates.

There are just a few exceptions to this rule, and the material provided is frequently of poor quality. Unless there are unusual conditions. Poor clinical procedures and a lack of laboratory access make obtaining test confirmation of some illnesses, such as Salmonella, challenging. Several nations have launched voluntary or non-mandatory food safety measures. In Thailand, training programs are utilized to improve food safety and hygiene in supermarkets, restaurants, and other foodservice organizations. Voluntary inspection and certification schemes are available to foodservice enterprises.

Despite this, industrialized countries continue to reject food produced in South Asia, according to research done by the International Food Policy Research Institute (IFPRI). South-East Asian nations are now collaborating with the World Health Organization to create a more comprehensive farm-to-table strategy.

6.8 DIVERSE SPECTRUM OF CUSTOMERS IN SOUTHEAST ASIA

In Southeast Asia, for example, consumer organizations have identified the development of food safety standards as one of the most critical objectives. Eastern Medicinal Region Food safety is a growing issue among people in the Eastern Mediterranean region. Countries in that region have conducted comprehensive audits of their food safety systems in collaboration with the World Health Organization, and some have updated their national legislation. Despite continual efforts to update rules, a lack of data prohibits a thorough evaluation of food safety hazards and obstacles (Gill et al., 2007).

6.8.1 Pathogens that can be eaten in the Eastern Mediterranean

Because many nations lack effective foodborne disease monitoring and reporting systems, it is hard to estimate how many individuals in the Eastern Mediterranean area get infected by food. Foodborne disease is a typical occurrence in this region, according to the World Health Organization and the Food and Agriculture Organization of the United Nations. Prescription medications can be obtained without a doctor's prescription, and diseases are not recorded. Waterborne infections, such as diarrhea and cholera, are sometimes thought to be more prevalent than foodborne illnesses. Before seeking medical assistance, many people wait until they have reached a point where only extreme therapies can help. Foodborne disease detection and diagnosis require considerable improvements.

Microbial contamination and the spread of foodborne infections can be exacerbated by a range of practices, such as eating raw and cooked salads and making raw milk cheese, as well as by specific food preparation processes, such as processing raw milk. Foodborne disease outbreaks have grown as a result of young people's greater proclivity to dine out and consume prepared meals. Despite this, foodborne illness appears to be declining in Gulf nations such as Saudi Arabia and Oman, despite difficulties in gathering reliable foodborne disease monitoring data. Milk pasteurization, food canning, herd immunization, economic expansion, refrigerator use, improved water supply, food monitoring, and more consumer understanding were all likely contributors to the rise in foodborne disease (Fung et al., 2018).

6.8.2 Policies and Plans for Action in the Eastern Mediterranean Region

Agricultural land is rare in several Eastern Mediterranean countries. Imports of food are critical to meeting consumer demand and ensuring food security in the United States. As a result, methods to ensure public health safety and quality are required. Food exports, on the other hand, are an important source of foreign money for neighboring Eastern Mediterranean nations (particularly those with non-oil economies). As a result, increasing food and agriculture export market share needs tight food safety processes. Outbreaks of Brucellosis, Rabies, and Salmonellosis in the Eastern Mediterranean Region have prompted the region to strengthen its food safety procedures. In a vast number of nations throughout the world, the construction of systems for the quick exchange of disease-related data, as well as technology cooperation agreements, control measures, and legislation have begun. However, due to a lack of cooperation across the public health, veterinary, food safety, and animal commerce sectors, preventing and managing zoonoses (infectious illnesses transmitted from animals to people) is challenging.

The great majority of food control regimes in the Eastern Mediterranean are incapable of dealing with new concerns. Foodborne disease in the United States is exacerbated by a lack of training in proper food handling methods, outdated rules, and uneven standards. Generally, the procedures utilized to test food are out of date, and the inspectors themselves are inexperienced in risk-based approaches to assuring food safety. Laboratories struggle to access the essential reagents or reference materials, and they lack worldwide recognition due to a lack of skill and funding. Governments and municipalities are also staunchly opposed to change. Many Eastern Mediterranean nations lack robust foodborne disease reporting systems that are in sync with national food control authorities (Fritsche, 2018). Even in countries with established reporting mechanisms, there is rampant underreporting. As a result, while creating national food safety programs, the frequency of foodborne illnesses is frequently overestimated, and chemical or microbiological pollutants are seldom considered. Food safety is supervised by several authorities in most countries. In Lebanon, for example, six different government bodies are in charge of food safety.

However, there is no comprehensive food safety regulation in existence, and the standards that are in place are not being implemented effectively. The European Union has banned some exports due to that country's excessive use of pesticides, hormones, chemical fertilizers, and antibiotics. Food hygiene and sanitation, as well as foodborne disease surveillance, are frequently handled by government health departments. Agricultural ministries are frequently in charge of food production, processing, and distribution, as well as the quality and safety of animal-derived foods. Because of the lack of agriculture in their economies and the concentration of food-related industries in urban areas, city administrations in the Gulf States are in charge of food management. Food safety is a top issue in the Eastern Mediterranean region, and in 1999, a regional plan of action to address food safety in the twenty-first century was approved. This critical resolution has spurred some countries to create food safety national profiles as well as thorough analyses of their food safety systems. Food regulations in Morocco and Tunisia, as well as several other nations, have been revised to match international standards. Morocco's five-year "Road Map" for unifying the country's food control system includes the creation of a food regulatory organization.

As a result of this plan, a scientific committee inside this institution will be in charge of assessing the country's food safety issues. Iran, Sudan, and Syria have all modified their food regulations and standards. Yemen's national profile on food safety has been finalized as of today. Egypt, Jordan, Morocco, and Tunisia have all adopted a risk-based food safety strategy. Jordan established a Food and Drug Administration to safeguard the safety of the country's food supply. The UAE has depended on advanced software for food inspection to assure the safety of food, both domestically produced and imported (Ergönül, 2013).

The Gulf Cooperation Council (GCC) nations are also keen on integrating their food laws, regulations, and standards (GCC). Food produced in or imported into any GCC member state may be circulated while retaining the same high standards of inspection due to the alliance's shared food export laws. It was also granted permission to engage with the Pan American Institute for Food Protection and Zoonoses on the establishment of food control in the Eastern Mediterranean Region. The principles can be used by governments to create food safety legislation that combines the need to safeguard public health with the requirement to satisfy internationally recognized standards.

In the Eastern Mediterranean area, good manufacturing practices (GMP), good agricultural practices (GAP), and Hazard Analysis Critical Control Points (HACCP) are becoming increasingly popular, as a program to educate farmers and other producers about their relevance. GAP has been advocated in Sudan, for example, with a focus on pesticide and fertilizer safety. Tunisia's fisheries sector is expected to adhere to HACCP standards as part of the country's food safety rules. Organic manufacturing is also gaining popularity as a means of reducing the probability of chemicals being overused or misused. The most challenging aspects of this approach are accreditation and acknowledgement.

Several Eastern Mediterranean enterprises have also adopted HACCP to improve food safety at home and gain export market share.

Consumers in the Eastern Mediterranean

Consumer organizations in the Eastern Mediterranean are worried about food quality, pesticides, food safety education, and awareness.

6.9 AFRICAN REGION

Food safety is a problem in Africa, as indicated by the high prevalence of diarrheal diseases in neonates and young children. Even though foodborne disease outbreaks are widespread in Africa, national governments have made little headway in building foodborne illness surveillance systems. Food contamination surveillance that is insufficient or nonexistent jeopardizes governments' capacity to appropriately estimate the impact of food contamination concerns on public health (EFSA et al., 2020). While poverty is the primary cause of hazardous food consumption in Africa, other factors such as a lack of clean water, insufficient government structures, population growth, the spread of AIDS and other infectious diseases, economic pressure, and poor environmental conditions exacerbate the problem. Overwhelming national limitations and a scarcity of resources make it impossible for Africa to evaluate the quality of imported commodities, exacerbating the continent's challenges.

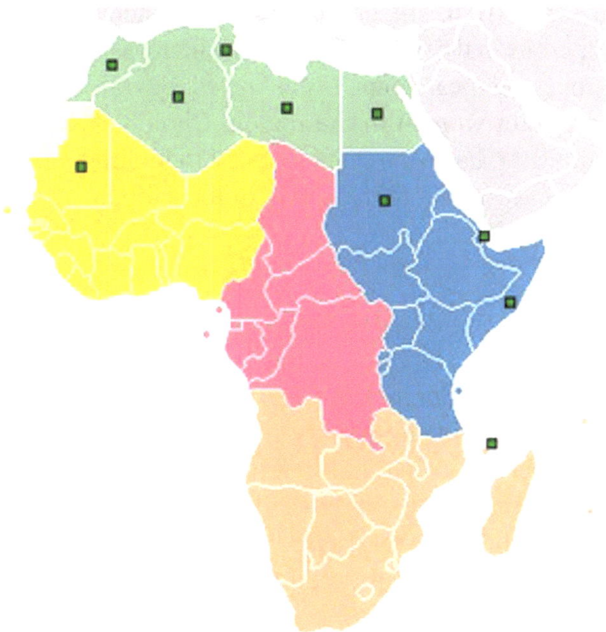

Figure 6.11: African region.

Source: https://upload.wikimedia.org/wikipedia/commons/thumb/c/cf/Carte_Afrique_CAF.png/290px-Carte_Afrique_CAF.png

In Africa, infectious diseases are spread via food.

Despite a lack of data on foodborne infections in Africa, research has revealed the presence of the following organisms: Campylobacter, Salmonella, Shigella, Hepatitis, Brucella, *Staphylococcus aureus*, *Bacillus cereus*, *Escherichia coli*, and rotavirus. Each year, an estimated 800,000 African children die as a direct result of diarrhea and dehydration. For example, in Zimbabwe, the incidence of diarrhea bouts lasting more than 14 days has been estimated to be as high as 6.05 percent among young children. Pesticide exposure may potentially have a role in the development of immunological and endocrine disorders, neurological illnesses, and cancer in African children (Diasolua et al., 2002). The number of people in this section of the country who are particularly vulnerable to foodborne disease is growing. Bacterial infections such as Salmonella can be fatal in Sub-Saharan Africa, where more than 25 million adults and children are infected with HIV/AIDS. Infections with enterohemorrhagic *E. coli* are especially dangerous in the elderly. Due to the increased use of foods such as refrigerated processed meat items, *Listeria monocytogenes* is also becoming an issue for pregnant women in the region. Cholera, which was formerly associated with water, has been demonstrated to be foodborne and is native to Africa.Africans worry about the safety of their food.

6.9.1 Poverty

While a variety of interconnected factors contribute to Africa's high prevalence of foodborne infections, poverty is the major culprit, disproportionately impacting women and children. Poverty exacerbates food safety concerns in a variety of ways, including unsanitary conditions in rapidly growing urban areas, a lack of access to safe drinking water, unsanitary food transportation and storage, and low levels of education among consumers and food handlers, resulting in limited food safety knowledge. Furthermore, national governments lack the financial resources required to expand foodborne disease surveillance and monitoring capability, fully execute food safety legislation, invest in modern buildings and utilities, and develop food safety education programs.

As a result, the World Health Organization (WHO) developed an integrated approach that integrates food safety concepts with national poverty reduction efforts.

Food found across the streets Street vendors is an excellent source of low-cost meals. However, because of inadequate regulatory systems, insufficient food safety rules, a lack of financial resources to invest in safer equipment, and a lack of food handler education, street meals often fail to meet basic hygiene criteria (Cody et al., 2012). Because street food is frequently prepared ahead of time for eating, it is susceptible to contamination by dust and insects. Furthermore, food workers may be infected with tuberculosis, typhoid, or other infectious diseases that can contaminate food. FAO and WHO have developed several initiatives aimed at improving the quality and safety of African street food. A program in South Africa, for example, instructs vendors and handlers on proper food preparation and handling practices. Guinea Bissau has funded a study to develop practical methods for improving the quality and safety of street meals, protecting clients, and reforming the street food business.

6.9.2 Mycotoxins

Mycotoxins, which are naturally occurring fungal poisons, pose significant food safety concerns. Aflatoxins are mycotoxins that have a significant impact on African public health. Mycotoxins contaminate a wide range of agricultural products both before and after harvest. Tropical conditions, such as high temperatures and humidity, monsoons, heavy rains during harvest, and flash floods, all promote fungal growth and mycotoxin production. Inadequate harvesting methods, insufficient storage, and less-than-ideal transportation and marketing conditions can all promote fungal growth and increase the risk of mycotoxin formation. AflatoxinM1 was found in human breast milk samples from Ghana, Nigeria, Sierra Leone, and Sudan, as well as umbilical cord blood samples from Ghana, Kenya, Nigeria, and Sierra Leone, indicating that it is a persistent contaminant in diets (Chen et al., 2020). Together with the hepatitis B virus, aflatoxins contribute to the increased risk of primary liver cancer in tropical Africa. Recent research in West African countries like Benin, Gambia, and Togo suggests that certain demographic groups and fetuses are chronically exposed to aflatoxins through their diets. Furthermore, children exposed to aflatoxins may experience delayed growth or chronic underweight, making them more vulnerable to infectious diseases during childhood and adolescence.

6.9.3 Food safety scares

In recent decades, the frequency and magnitude of humanitarian crises in the African Region have had a significant impact on food safety. Food supplies are often destroyed or considerably damaged in the aftermath of natural disasters such as floods, droughts, and earthquakes, as well as harsh civil wars or border conflicts, wreaking havoc on survivors' health. Foodborne disease outbreaks are common in refugee camps as a result of filthy living conditions, environmental contaminants, and improper food handling. A massive cholera outbreak plagued Rwandese refugee camps in Goma, Zaire (the Democratic Republic of the Congo) in 1994, resulting in an estimated 70,000 cases of diarrheal disease (mainly cholera) and a substantial death rate. In 1992, 772 cases of stomach aches and bloody diarrhea were documented in Malawi's Lisungwi camp, which housed 60,000 Mozambican migrants.

6.10 THE ECONOMIC IMPACT OF FOODBORNE INFECTIONS

Foodborne illnesses have a wide range of negative economic consequences in Africa. A cholera pandemic in Tanzania, for example, cost $36 million in 1998. Nigeria's Food and Drug Administration disposed of aflatoxin-contaminated food valued at more than US $200,000. The World Health Organization has highlighted several food safety and quality issues affecting African food exports and imports. Spoilage, substandard/fake products, failure to provide production dates, improper or deceptive labeling of imported food, poor product quality and packaging of exported food, expired food, exceeding preservative/additive levels, lack of harmonization of food safety regulations, and fraud are all examples (Chandrapala et al., 2012).

African countries frequently lack adequate inspection and laboratory capacity, and their exports may be barred if microbiological and other standards are not met. These concerns have frequently resulted in the importation of subpar food products and the rejection of food exports by foreign markets. Export rejection losses cost nations not just tremendous financial resources, but also their credibility as reliable trading partners. Ugandan fish exports to the European Union (EU) were prohibited in 1997, for example, since the country's fish processors and exporters failed to meet the EU's new cleanliness and processing quality requirements. Uganda lost $36.9 million in income as a result of lower returns during the prohibition, which ended in July 1999. Furthermore, the fishing industry loses around $720,000 in monthly profits.

6.11 AFRICAN POLICIES AND ACTION PLANS

Because many African countries lack appropriate food supplies, developing efficient food control systems is a secondary concern. Often, marketing, implementing, and enforcing dietary restrictions receive less attention.

6.11.1 Food Laws, Regulations, and the Food Safety and Inspection Service

Basic food rules may not be integrated into African regional legislation, or they may be out of date, fragmented, or just insufficient. The legal structure might be perplexing at times for enforcement officials, manufacturers, and distributors. Numerous ministries or departments are involved in food safety processes, resulting in duplication of efforts, resource duplication, and enforcement gaps. When it comes to food control policies, it may be difficult to determine which agency represents which country. Nonetheless, great progress has been made in this area. In 2004, Madagascar established a single food safety organization known as the "Unité de Contrôle de Qualité des DenréesAlimentaires." Food control systems in Africa are mostly based on colonial-era laws. Many countries have not updated these institutions, which were crudely designed to deal with issues of particular relevance to colonial rulers (Carvalho, 2006). The majority of African countries have attempted to modernize their outmoded food legislation. In Mauritius, for example, a new Food Legislation was established in June 1998 (to replace the 1940-era statute) and went into effect in January 2000. The food industry, on the other hand, has criticized the law for failing to meet global standards.

6.11.2 Services in Surveillance, Laboratory Testing, and Food Inspection

Disorganized sampling is common in Africa due to a lack of coordination among surveillance, food laboratory, and food inspection institutions. Furthermore, the emphasis is on sampling for enforcement purposes, and food pollutants are seldom monitored systematically. Inadequate recordkeeping can set off a vicious cycle, resulting in a scarcity of data on which to base local policy, law, and dietary standards. Furthermore, only a few nations have monitoring systems capable of detecting frequent foodborne illness microorganisms. As a result, surveillance data is inconsistent and unreliable. In conjunction with the Pasteur Institute in France, WHO has been providing training on foodborne illness surveillance and microbiological monitoring of foods for Francophone nations in the African Region. Training programs have yet to begin in Anglophone and Portuguese-speaking nations. Food safety control laboratories frequently operate inefficiently owing to a lack of financial resources for equipment development and maintenance. Furthermore, the great majority of African public health laboratories are incapable of screening for chemical pollutants and naturally occurring poisons. Food inspectors must be informed about food safety and related concerns, which appear to be missing in most of Africa. For example, the majority of the region's countries lack specialized mechanisms for collecting and sharing data on food exports that have been rejected by international purchasers. Food safety regulation is further hampered by a lack of competent inspectors (the Comoros and Mauritania each had only two food inspectors in 2002) and a lack of cooperation among key agencies.

6.11.2 Education and Training

Food safety education for companies and consumers is scarce in the majority of African countries.

Furthermore, there is a global dearth of qualified professionals to support laboratory services, and the majority of African nations lack the essential infrastructure to develop local expertise.

Funding for food safety initiatives is sometimes limited since many African governments consider they have more vital obligations. Due to a lack of resources at the local and national levels, regional activity is carried out with the aid of international organizations. For example, the Common Market for Eastern and Southern Africa (COMESA) has collaborated with the FAO to create food safety and quality control systems while simultaneously fostering policy harmonization. Africa also features a variety of consumer organizations (Beulens et al., 2005).

Consumer groups in the African Region, such as those in Senegal, Nigeria, Cameroon, and Benin, are very active in food safety initiatives and regularly take on activities that were previously done by government organizations in other regions. Certain organizations oversee food safety and market and shop inspections; conduct chemical, bacteriological, and physical examinations of food goods; monitor recalls of contaminated foods, and guarantee that the government or industry notifies consumers. In other places, the emphasis is on boosting both the quality and awareness of street food, as well as consumer education and awareness. In Benin, one group is building a "Customers House" to monitor imported commodities and inform consumers on the quality of imported food supplies.

6.12 EUROPEAN REGION

Following the BSE catastrophe in the United Kingdom and the dioxin contamination of animal feed, which affected a range of meat products in the continent's north, Europe's food policy was amended. The EU (European Union), which consists of European member states, established an integrated approach to food safety and a set of food safety standards to harmonies existing national legislation. EU food safety rules are continually inventive since they are still in their infancy.

Figure 6.12: European Region.

https://www.worldatlas.com/r/w768/upload/a6/8b/e0/countries-of-southern-europe-1.png

6.12.1 Foodborne Illnesses in the European Union

According to a WHO study, the prevalence of foodborne diseases in Europe is increasing, despite difficulties in assessing their entire impact. Since 1985, the number of *Campylobacter jejuni* infections has steadily climbed. Perhaps the apparent increase is due to improved diagnosis rather than a higher frequency in some countries. Campylobacteriosis is a type of spontaneous illness that can affect a single person or a small group of individuals. Despite a drop in the prevalence of the disease in some countries, Salmonella remains the most often reported cause of foodborne outbreaks. The most prevalent cause of outbreaks, *Salmonella enteritidis*, accounts for 75% of all cases (Borchers et al., 2010). Some non-Muslims in the Balkans are suffering from Trichinellosis, a parasitic infection that is becoming more common as a result of the consumption of homemade pork products that do not follow basic veterinary requirements. Since the mid-1990s, certain infections, such as brucellosis (Malta fever), have been progressively increasing in Central Asian nations.

The predominant method of transmission for Brucellosis in these countries is raw goat and sheep milk. The surge is being blamed on socioeconomic and political concerns that limit livestock management efforts, as well as a lack of understanding about the sickness. Traditional food preservation procedures have resulted in a high frequency of botulism in Eastern Europe.

The Caucasus is said to have the highest botulism rate in the world. – Wiktionary (Armenia, Azerbaijan, and Georgia). Because fresh vegetables are expensive and canned food is scarce in these countries, traditional canning methods are commonly employed. Pesticides and other pollutants have been discovered in a wide range of foods, including fish. As a result of prolonged exposure, many of these compounds have a harmful influence on the endocrine and respiratory systems, as well as the development of cancer. Industrial pollution, such as mining and smelting, energy production, agriculture, and the transportation of hazardous and municipal waste, is the leading cause of food poisoning in Central and Eastern Europe. To mention a few, the Aral and Caspian Seas are examples. Food tampering, whether accidental or deliberate, has had serious consequences for public health. Many people have been permanently injured by aniline-denatured rapeseed oil, which has killed around 2,000 people and rendered another 20,000 disabled (Antle, 2001). Despite a thorough inquiry, the source of the contamination was never discovered. There are just a few countries where *Listeria monocytogenes* outbreaks have been documented, with France being the most prevalent.

CHAPTER 7

FOODBORNE ILLNESS AND THEIR CAUSES

CONTENTS

7.1 INTRODUCTION

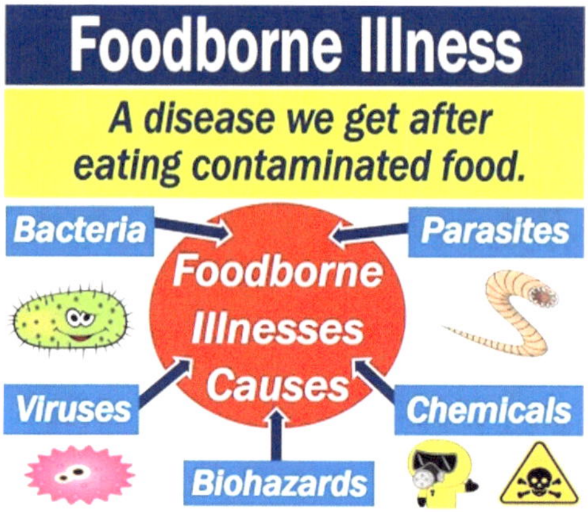

Figure 7.1: What is foodborne illness?

Source: *https://i0.wp.com/marketbusinessnews.com/wp-content/uploads/2018/06/Foodborne-Illness.jpg?resize=406%2C346*

The contamination of fresh fruit and vegetables or packaged foods with virus-containing fecal material is the source of many foodborne diseases, despite the fact that viruses, unlike bacteria, rarely thrive on or in foods. Noroviruses, hepatitis A and E viruses, rotaviruses, and astroviruses are the most usually reported major foodborne diseases. Viruses have been blamed for as much as 67% of all foodborne infections, and a rise in the transmission of viruses via food and water has lately been recognized. Immunization against such pathogens has long been acknowledged as the optimal way to give protection against sickness and disease outbreaks connected with these viruses because of their highly infectious nature and ability to survive under harsh circumstances, such a high acidic pH or cold temperatures (Bouzembrak et al., 2019). Prophylactic vaccinations against acute viral infection related to food and clean water in people of all ages have recently been developed as a result of an expanded understanding of their clinical importance and effect. There are currently no vaccinations for foodborne viruses other than hepatitis A, which is the only one now accessible. Hepatitis A outbreaks have been greatly decreased as a result of the widespread vaccination of certain individuals at risk. Vaccines against noroviruses as well as rotaviruses might considerably lower the incidence of foodborne disease associated with these viruses in the near future, based on current research and literature.

Foodborne sickness is most often caused by inappropriate handling, preparation, or storage of food products. Preventing sickness by following good hygiene standards prior, while, and even after food preparation may help to lower the likelihood of developing an infection. In the public, there is general agreement that frequent washing hands is among the most efficient defense against the transmission of foodborne diseases and other illnesses. Food safety refers to the process of inspecting food to verify that it does not contain any pathogens that might cause foodborne diseases in humans. Foodborne illness may be caused by a wide range of chemicals that have an impact on the environment. Furthermore, a foodborne disease may be caused by a variety of chemicals, including pesticides, medications, and naturally occurring toxic compounds, such as vomitoxin, deadly mushrooms, and reef fish, among others.

Any ailment produced by the intake of contaminated foods or drinks, whether infectious or noninfectious, is known as a foodborne disease or foodborne illness. Bacteria, viruses, and parasites are the most common causes of foodborne disease. Besides mycotoxins, marine biotoxins, and the contaminants found in poisonous mushrooms, other agents include metals like lead and mercury; organic pollutants like dioxin, which seem to be byproducts of some manufacturing processes; and prions, which can be transmitted by eating contaminated food (Boudergue et al., 2009). In addition to gastroenteritis and reproductive and developmental problems as well as neurological diseases like bovine spongiform encephalopathy, foodborne illness agents may cause a wide range of other health problems. Food poisoning is a term that is used to describe some types of foodborne infections, primarily gastroenteritis. Although it is classified as a foodborne disease, food poisoning may only be contracted when bacteria generate toxins in the food that cause symptoms to appear shortly after consumption. Other foodborne diseases, on the other hand, might develop over the course of many days after ingesting the harmful microorganisms or chemicals.

The burden of foodborne disease is estimated to be significant, despite the fact that many instances go unreported. Diarrhea is a popular backup method for foodborne illness, with tens of millions of individuals impacted each year. Outbreaks of foodborne diseases are common, and they may impact a significant number of individuals. More than 300,000 individuals were infected with hepatitis A in China in 1988 after eating contaminated clams, while 224,000 people in the United States with salmonellosis in 1994 after eating tainted ice cream. It is possible to die from a foodborne infection.

Out of 142 people infected with Listeriosis, 48 died as a direct result of an incident in California in 1985, which was linked to a tainted cheese product.

The general causes of foodborne diseases can be classified under the following groups: Bacteria, pathogens, and parasites are examples of biological risks. Most foodborne infections are caused by bacteria and viruses. The greatest danger to food safety is posed by biological agents. The product may have intrinsic flaws, or they may have been mishandled. Biological poisons and chemical pollutants are examples of chemical risks. Foodborne bacteria may produce natural poisons when they infect food that has been overcooked or undercooked. It is possible for certain persons to be poisoned by certain additives, such as sulfites. When items are utilized incorrectly, chemical contamination might develop. Allergens in food pose a chemical threat. Certain foods include proteins that might cause allergic reactions in certain individuals. It's impossible to generalize about food since every dish is unique. Among the most common dietary allergies include fish, wheat, and soy. Plastic and metal fragments from cans and shattered glass are examples of physical risks.

7.2 BIOLOGICAL CAUSES OF FOODBORNE ILLNESS

There is little doubt that biological pollutants are the main cause of sickness. The remainder of the book's techniques for food hygiene and safety and prevention are crucial in reducing or eliminating many of the problems connected with biological contamination. We are always surrounded by microbes. If you don't have a microscope, you can't see them at all. While certain microorganisms might be useful, others can be harmful and even fatal. They are referred to as pathogens when they cause disease (Baluka & Rumbeiha, 2016). Bacteria may be found in a wide variety of foods, as well as in the human body. Some bacteria are hazardous and may cause sickness, while most bacteria are harmless and even useful to humans. Examples of pathogenic bacteria include Campylobacter, *E. coli*, Listeria, and Salmonella. It is imperative that food containing these germs be handled and cooked properly. The presence of viruses in food does not increase their ability to cause sickness, but rather their ability to remain dormant.

People handling food are responsible for most cases of foodborne disease caused by viruses because of inadequate food handling or sanitation practices. Viruses such as Hepatitis A and Norovirus are common causes of food poisoning. Food infested with parasites may cause disease if it is not prepared to a high enough temperature or refrigerated to a low enough temperature to destroy the parasite. One-celled creatures known as protozoa may very well be found in groundwater. Consumption of water that has been contaminated might result in disease. When organic stuff decomposes, fungi thrive. Molds that develop on ruined food, for example, might be toxic long after you cut or scrape the apparent mold off the meal. Many fungi are useful or innocuous.

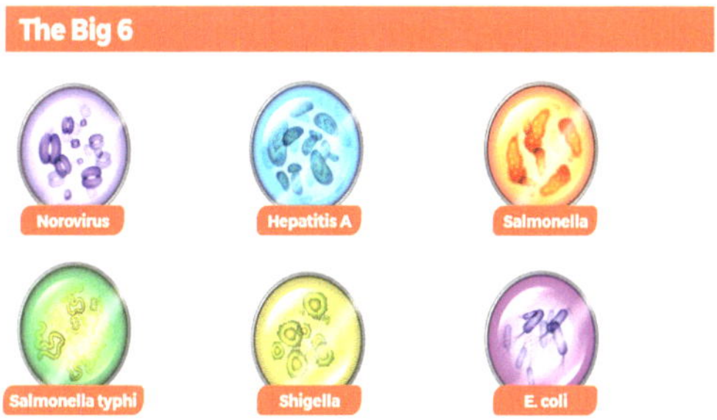

Figure 7.2: Examples of foodborne illnesses.

Source: https://alwaysfoodsafe.com/images/uploaded/87368307_3546747458699995_6928252925324034048_n.png

Examples of the foodborne illnesses and their causes include:

7.2.1 Norovirus

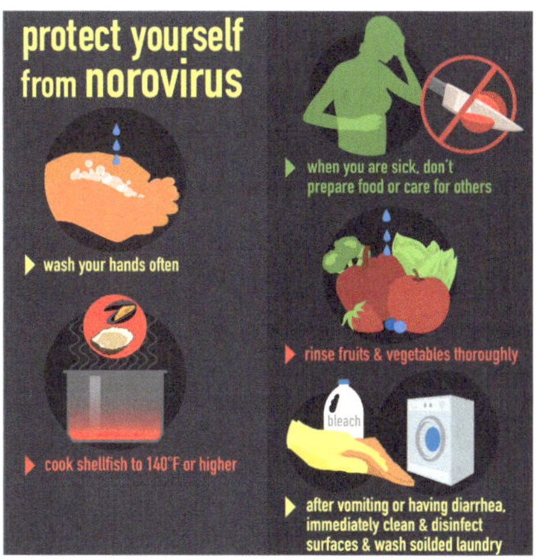

Figure 7.3: Norovirus as a foodborne illness.

Source: https://www.cdc.gov/norovirus/images/info-norovirus-poster.png

Gastroenteritis is the second most often reported condition in the United States, just behind respiratory infections as a reason for people seeking medical attention. The World Health Organization estimates that one billion episodes of acute diarrhea occur each year in children and adults. – It is rather uncommon for gastroenteritis to present itself as a minor case of diarrhea, but it may also present itself as a more severe case, resulting in dehydration and death. Gastroenteritis is thought to be the cause of death in between four and six million individuals each year. A wide range of pathogens may cause diarrhea, including viruses, bacteria and parasites (Zanin et al., 2017). In impoverished nations, bacterial agents are more prevalent, while in developed countries, viral pathogens are more prevalent. As far as public health and sanitation are concerned, these agents are critical. There were fecal samples from a gastroenteritis epidemic in Norwalk, Ohio, in 1972 when 27-nm virus particles were found in an infectious filtrate. Currently, four families of viruses are thought to be responsible for the majority of gastroenteritis in children.

When it pertains to gastrointestinal sickness, noroviruses (NoVs) have long been regarded the second most common cause of gastroenteritis, behind only the more well-known rotaviruses. The improvement of molecular tools in the diagnosis of NoV has allowed a greater perspective into the epidemiological effect of these viruses, which are presently identified not only as of the largest reasons for non-gastroenteritis epidemics, but also as a significant source of sporadic gastroenteritis in both children and young adults. The norovirus virus is very infectious. Infection with norovirus results in gastroenteritis which is considered to be inflammation of the stomach and intestines. This results in nausea, diarrhea, and cramps in the stomach. In addition to food poisoning as well as stomach flu, norovirus infection is also known as gastroenteritis. Other microorganisms and chemicals, as well as noroviruses, may cause food poisoning. The flu and norovirus are not connected in any way (influenza). A respiratory ailment, influenza is caused by the influenza virus, despite the fact that some of the symptoms are similar.

There are several ways in which a person might get infected with norovirus. In order to get the virus, you must come in direct contact with someone that is ill, touching infected surfaces, and then touch your lips. In the United States, norovirus is the most common cause of foodborne disease. Whenever an infected food handler handles foodstuff before serving it, the virus is frequently disseminated. Oysters and other shellfish may be naturally infected with norovirus, which can cause diarrhea.

Vomiting, non-bloody diarrhea, stomach pains, and biliousness are among the symptoms reported by the Centers for Disease Control and Prevention. These seem to be the result of a 24-48-hour development period, however, there are occasions when indications appear just 12 hours after exposureto the virus has followed. Norovirus disease frees the large intestine, resulting in the absence of fecal leucocytes in stool samples from stool testers, according to a study of grown-up volunteers who were injected with the virus (Xiang et al., 2019).

Infected persons' feces and vomitus contain noroviruses. Viruses like this one may spread quickly across healthcare facilities because they are so easily transmitted. There are various ways in which a person might get infected with the virus: Infected individuals may be spread by direct contact with infected individuals (e.g., healthcare workers, guests, or other patients). Consuming norovirus-infected food or beverages is a cause. Norovirus may be spread by touching the skin or things that have been contaminated with the virus.

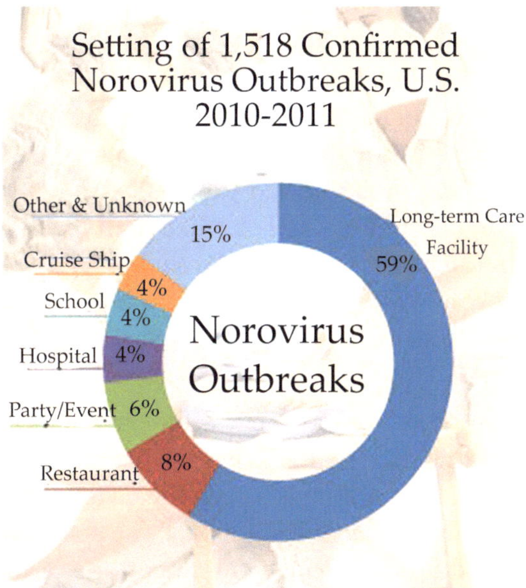

Figure 7.4: Statistical Analysis of Norovirus outbreak.

Source:https://www.foodsafetynews.com/files/2013/01/dsnorovirus_379px-1.jpg-1.jpg

7.2.2 Norovirus Prevention

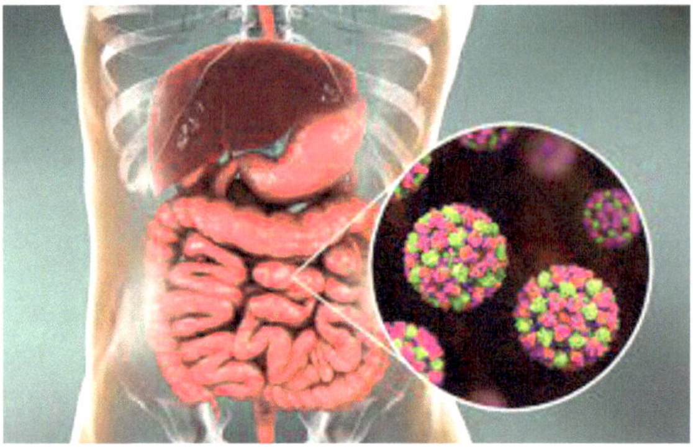

Figure 7.5: Preventive measures against Norovirus.

Source: image/jpeg;base64,/9j/4AAQSkZJRgABAQAAAQABAAD/

Patients who are suspected of having norovirus may be housed in isolation rooms or shared with other patients who have the same illness. Healthcare facilities should take additional measures to prevent noroviruses from spreading: Wash your hands thoroughly with soap and water after coming into contact with somebody who has been exposed to norovirus. When working with norovirus-infected patients, wear gowns and gloves to protect yourself. Regularly clean and sterilize high-touch patient exteriors and equipment using an Environmental Protection Agency-approved product with a tag claim for norovirus, remove and wash contaminated clothes or linens, healthcare workers with indications consistent with norovirus should be banned from work.

Person-to-person contact, contact with contaminated fomites, or consumption of feces-tainted food or drink are the most common routes by which the virus is spread. Vomiting has been linked to airborne transmission in recent epidemics at educational institutions. The spread of noroviruses at a facility may be sped up notwithstanding the efforts of the facility's staff to contain the spread of the disease. The Bureau of Infectious Diseases, Division of Public Health has compiled a list of "recommended, but not mandatory" control measures based on information from previous hospital eruptions to assist in the decision-making process for facilities and nursing homes when similar contaminations occur (Wilcock et al., 2004). Depending on the number of patients and the degree of dispersion inside the institution, the motions taken by various hospitals or nurturing homes may differ.

Patients with a Norovirus infection are extremely sensitive because of the unpleasant symptoms of this illness. Personal attention from nurses is essential for patients with mental health issues. For sanitary reasons, nurses must pay close attention to each patient. Nurses must pay special attention to patients who are receiving symptomatic therapy. Nurses should wear gowns, gloves, and masks at all times and should not leave the ward often to avoid the transmission of the illness.

7.3 SALMONELLA

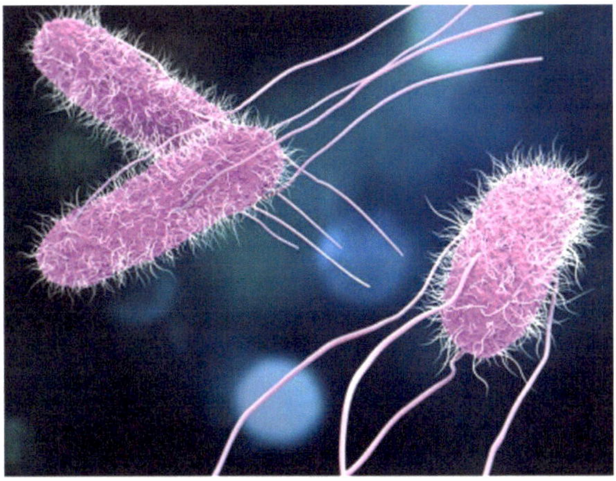

Figure 7.6: Salmonella Disease.

Source: https://www.fda.gov/files/salmonella_salmonellosis.jpg

Getting sick with salmonella is very common, and it usually happens to people who eat food that has been contaminated with bacteria called salmonella. When salmonella bacteria survive in the digestive tracts of animals and humans, they are excreted in their waste. The most common way for humans to get infected is via dirty food or drink.

It's very common for people who get salmonella to not show any signs of being sick. There are also some people who have diarrhea for 8 to 72 hours. They also have a fever as well as abdominal cramps. Recovery time for the majority of healthy persons is just a few days. Salmonella infection may cause dehydration in certain circumstances, necessitating quick medical intervention (Wertheim-Heck et al., 2019). Infections can sometimes be life-threatening if they spread to other parts of your body, such as your brain or spinal cord. If you go to nations with poor sanitation, your chance of contracting salmonella increases. Salmonella infection is mainly caused by the consumption of raw or undercooked meat, poultry and their products. It can take a few hours to 2 days for the embryo to grow into a full-grown adult organism. Most salmonella infections may be classed as stomach flu which is scientifically known as gastroenteritis. Salmonella clinical manifestations usually last for two to seven days, but they can last longer or shorter. Diarrhea might continue for up to ten days, but it can take many months for bowel function to recover to normal. Certain strains of salmonella bacteria cause typhoid fever, a potentially fatal illness that is much more prevalent in underdeveloped nations.

7.3.1 Causes

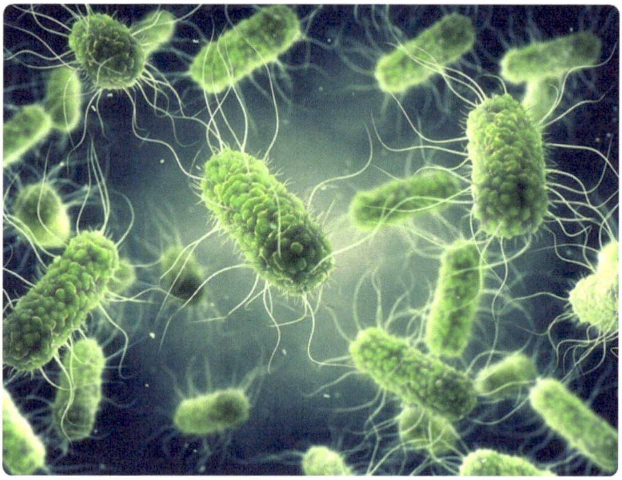

Figure 7.7: Causes of Salmonella.

Source: https://www.news-medical.net/image.axd?picture=2021%2F2%2Fshutterstock_1542887327.jpg

Salmonella bacteria dwell in the guts of humans, animals and birds. Salmonella is almost often acquired by the consumption of food that has been tainted by excrement. Commonly affected foods include:

- Raw meat, poultry and shellfish. Feces may get into raw meat and fowl during the slaughtering process. If the water in which the seafood is gathered is polluted, it is possible that the meal will be affected (Wahyuni et al., 2019).

- Raw eggs. While an egg's shell may appear to be a perfect barrier to infection, some sick hens produce eggs that carry salmonella before the shell is ever formed. Raw eggs are often used to make mayonnaise as well as hollandaise sauce at home, but they are not cooked.
- Fruits and veggies. The water used to hydrate fresh produce, opportunities for growth varieties, may be contaminated with salmonella. This can happen in the field or when it is washed throughout processing. Contamination may also occur in the kitchen when raw meat and poultry fluids come in contact with raw ingredients such as salads.

Additionally, according to the Food and Drug Administration, certain salmonella outbreaks have been linked to contamination in spices. Spices are now under investigation by the agency for methods to make them safer. After using the bathroom or attempting to change a diaper, many people don't wash their hands very well. When they cook, several food products get dirty. Infection also may develop if you touch anything that is infected, including pets, particularly vertebrates, and afterwards place your hands in your mouth (Tirado et al., 2010).

USDA has developed a Salmonella Action Plan that includes modernizing the system for inspecting poultry slaughterhouses as well as boosting sampling and testing procedures for poultry as well as meat products in general. People in the United States get sick from salmonella all the time. This plan is meant to cut that number down. You may also take precautions to not infect others with microorganisms. Preventive approaches are particularly crucial when preparing meals or giving care to newborns, elderly persons and those with compromised immune systems. Make sure you cook your food properly and store or freeze it right away. In order to keep salmonella bacteria from getting into one's mouth anywhere on any meal you're cooking, make sure you clean your hands very well. Cookie batter, handmade ice cream even eggnog often include raw eggs. If you absolutely must ingest raw eggs, be certain they are pasteurized.

7.4 HEPATITIS A

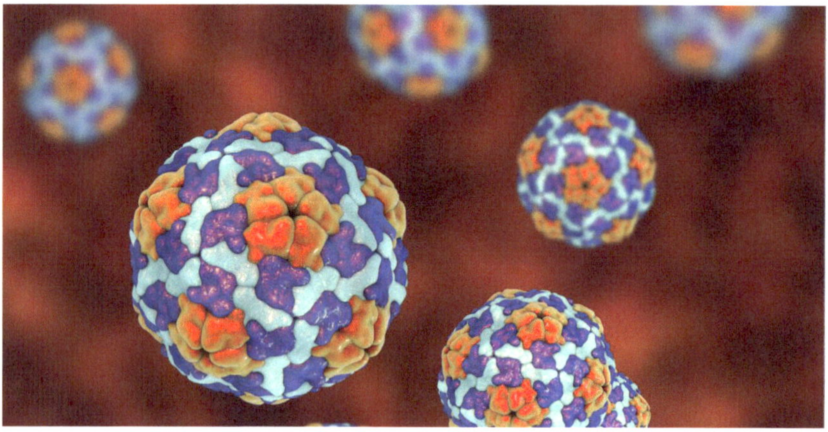

Figure 7.8: Causes and signs of Hepatitis A.

Source:https://hips.hearstapps.com/hmg-prod.s3.amazonaws.com/images/hepatitis-a-virus-illustration-royalty-free-illustration

An infection of the liver with the extremely infectious Hepatitis A virus causes Hepatitis A. The pathogen is one of the numerous forms of hepatitis viruses that causes inflammation and impairs the performance of the liver. Hepatitis A may be acquired by ingestion of tainted food or drink, or through direct contact with an infected person or item. Hepatitis A infections that are mild do not need treatment. The majority of persons who get the infection recover entirely, with no lasting liver impairment. One of the most effective strategies to prevent hepatitis A is to practice proper hygiene, especially regular hand washing. Vaccines are accessible tothose who are most vulnerable. When food is grown, harvested, processed, handled, or even cooked, the hepatitis A virus could get into the food. In areas where hepatitis A is widespread, contaminated supplies of food are more likely to be found. Because individuals in the United States eat a lot of imported food that isn't safe, there have been some foodborne epidemics in the country. The sickness is typically minor. It often begins between two and four weeks following consuming infected food or drink but also resolves within one to two weeks (Todt et al., 2009).

It's possible for people who are infected to not show any signs whatsoever, or even to be very sick. Approximately 30 percent of infants younger than 6 years of age have signs, whereas 70 percent of older kids as well as adults experience manifestations. In the event that an individual shows signs of being sick, they could be exhausted and not want to eat. People might also be nauseated and also have dark pee or jaundice, as well as fever, vomiting and diarrhea. As long as these signs persist, they could last for six months or more. In people who have hepatitis A virus, even though they don't show any signs, the virus still can spread.

In general, people who work with meals aren't more likely to get HAV due to their job. They're important since they play a big role throughout the spread of prevalent food-borne HAV. Kitchen staff makes up about 8 percent of the overall working population, which is about the same number of people who get hepatitis A seems to be those who handle food. Many food workers with hepatitis A do not transfer the infection to others. How the healthcare system responds to each case is based on a very meticulous risk evaluation? This may include on-site visits by local environmental personnel as well as jurisdictional consumer safety law enforcement agencies, self-governing verifying of sanitary conditions, as well as a comprehensive procedure as well as document analysis. The assessment process should figure out how many people might have been revealed and how likely they are to get a severe or non-serious illness, as well as how the media and the government might affect the decision.

If the person who handles food was sick and dealt with food that wasn't prepared before serving and also had digestive problems or poor hygiene practices, you might want to check up with patrons. Attendees should be recognized and given PEP in under two weeks after exposure. PEP should be addressed in institutional care when numerous patron interactions are possible. Threats toward other workers and guests must be managed together with food safety inspectors from the jurisdiction. PHU personnel must work very closely with these officers (Sinha, 2009). Customers as well as co-workers can get food poisoning from people handling food that has been infected. People who manage food could also get it from plates and cutlery or surfaces that are dirty. If the inquiry reveals that additional food workers at the establishment are at risk in the event of consuming meals made by the subject or sharing bathrooms or washrooms with the case, PEP should always be administered to those that are not immune. Blood must be taken from kitchen staff who are believed to cause other illnesses if they give their permission.

7.5 CAMPYLOBACTER

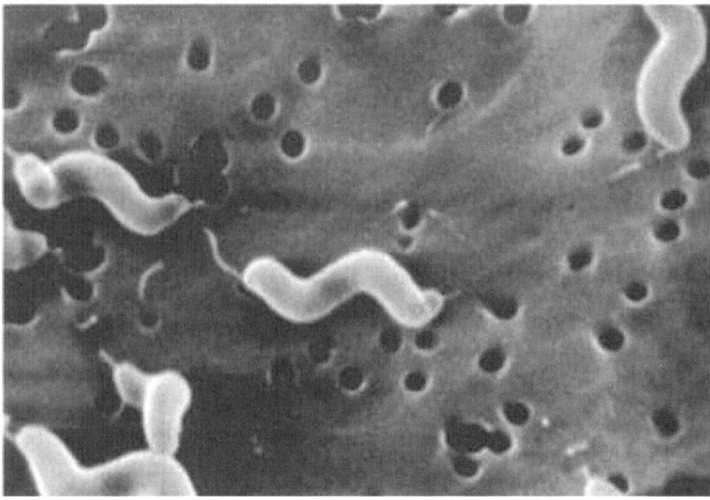

Figure 7.9: Causes of Campylobacter.

Source: https://img.medscapestatic.com/pi/meds/ckb/31/36931tn.jpg

The infectious disease caused by Campylobacter pathogensis known as Campylobacteriosis. It is the most often occurring bacterial source of diarrhea in the U. S. It is estimated that 20 incidences of food poisoning per 100,000 individuals are detected each year according to Food Net data. Twenty infections per 100,000 persons are detected every year by the Foodborne Infections Active Surveillance Network. Several more instances go unrecognized for fear of stigmatization. An estimated 1.5 million Americans have Campylobacter illness per year, according to the CDC. There are a lot more instances in the warmer months than in the winter, and they aren't linked to any known epidemic. Infections involving Campylobacter frequently cause diarrhea that seems to be bloody, as well as a fever and cramping in the stomach. There is a chance that the diarrhea could cause nausea and vomiting. It normally takes 2 to 5 days for the symptoms to appear following infection, and they often continue for one week or more (Sadiq & Beauchemin, 2017). Irritable bowel syndrome, transient paralysis, and arthritis are among the most possible side effects that these patients endure. People with compromised immune systems, including those with AIDS, a blood condition, or who are undergoing chemotherapy, are more susceptible to Campylobacter infection.

As much as one-third of all poultry as well as other livestock are infected with Campylobacter. Because of its ability to survive in many organs of livestock, Campylobacter may spread to certain other edible portions of a carcass during the process of butchering and preparation. Twenty-four percent of raw chicken purchased via stores tested positive for Campylobacter throughout 2015, according to NARMS testing. It is possible for milk to be contaminated by Campylobacter or dung from a cow that is infected in her mammary. Milk may be safely consumed after pasteurization. Soil as well as water polluted with cattle, bird, or even other animal excrement may infect produce, including fruits and veggies. There is a risk that animal excrement may pollute rivers and lakes. Consumption of contaminated water as well as untreated food may be minimized via proper washing and disinfection.

Only just few Campylobacter organisms are needed to cause illness from this bacterium. As little as a drop of liquid from a piece of raw chicken might infect an individual with germs. Foods that have been contaminated with Campylobacter are likely the source of most illnesses. Other items, such as fish, meat, fresh vegetables, as well as contact with infected animals and the consumption of unclean water, may also spread Campylobacter. Whenever raw or minimally cooked items, including salads or fruits, are prepared on a chopping board that had been previously used to chop as well as prep raw chicken, individuals are at risk of contracting foodborne illness. Exposure to dog or cat excrement may potentially lead to the spread of diseases. A donation of tainted blood has resulted in a small number of cases of infection. As a general rule, Campylobacter doesn't transfer from individual to individual (Smith & Pillai, 2004). Poultry, raw milk products, shellfish, tap water, fruit, and canines have all been linked to this particular epidemic. Individuals who go to poor nations are more likely to get campylobacter illness. Food Net reports that around about one five cases are linked to overseas travel of people with Campylobacter.

Whenever Campylobacter bacterium is found in feces, bodily tissues, or fluids, a diagnosis of this infection is made. A culture might be used to identify the bacterium, or a fast testing procedure could be used to look for the organism's genetic makeup. Most Campylobacter infections can be treated without the use of antibiotics. For as long as the diarrhea persists, individuals should keep up their fluid intake. Patients with and at risk of serious disease may need antibiotics. As you get older, you're more likely to get sick. You're also more likely to get sick when you're pregnant. A Campylobacter ailment usually goes away in less than a week, but some individuals continue to lose Campylobacter germs in their feces for days after they've recovered from the illness. Infections caused by Campylobacter are seldom harmful to the patient's health in the long run. Campylobacter infections can cause irritable bowel syndrome as well as arthritis in 5–20 percent of patients, according to certain research.

Guillain-Barré syndrome is a condition that occurs in around one out of each 1,000 cases of Campylobacter infection (GBS). GBS occurs as a result of an infection triggering a protective immune response in this way. Symptoms of GBS, such as muscular denervation, may linger for days and need extensive medical attention. Most individuals make a full recovery; however, a small number of people are left with lifelong neurological damage as a result of their injuries. It is estimated approximately 40percent of all GBS infections may be induced by Campylobacter infected persons living in the United States.

It is the USDA-job FSIS's to oversee the production and processing of meats as well as eggs. At least 10% of entire broiler poultry samples must be free of Campylobacter infection in processing industries, according to new guidelines imposed by the Food and Drug Administration in 2011. More recently, in 2016, USDA-FSIS mandated that raw chicken pieces be tested for Campylobacter infection. Additionally, the USDA-FSIS publishes the findings of its contamination tests. All food products not controlled by USDA-FSIS are subject to FDA oversight (Sargeant et al., 2009). Among the models FDA uses to regulate the food retailing industry are places like eateries, grocery shops, and even long-term care facilities. Legislation enacted by the Food and drug administration in 2011 called the Food Safety Modernization Act changes the focus from inhibiting bacterial growth to making sure that the food we eat in the United States is safe.

7.6 E. COLI

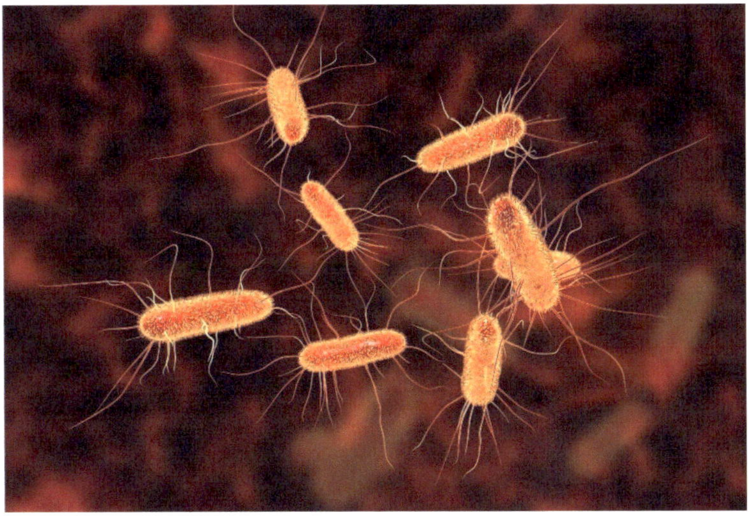

Figure 7.10: *E. coli* Bacteria.

Source: *https://familydoctor.org/wp-content/uploads/1999/05/96323925_l.jpg*

A huge and varied collection of microorganisms is known as *E. coli*. Despite the fact that the majority of *Escherichia coli* are safe, some may cause illness in humans. Shiga toxin produced by certain strains of *E. coli* is responsible for making people ill. The *E. coli* that produces this Shiga toxin which is usually abbreviated as STEC, is the name given to the microbes which produces these poisons. It is common for individuals to get infected from STEC in instances where they consume food that is contaminated with the bacterium. When it comes to STEC, there is a strain known as the O157 which is the most often documented strain throughout the United States. The term "non-O157" refers to STEC that do not contain O157. Hemolytic uremic syndrome, which is a rare but potentially deadly form of kidney failure, may be caused by certain strains of the bacteria STEC (Redmond & Griffith, 2004).

Escherichia coli is highly known for its adaptive nature which makes it one of the most adaptable bacteria known and has proven to be very dangerous. There is a technology called the Recombinant DNA technology which has made extensive use of *E. coli*, a natural component of the human beings and other animal intestinal microbiota. Aside from its role in the research as well as in the digestive tract, *E. coli* may potentially become an extremely dangerous virus. Virulence factors from a variety of *E. coli* variants may induce a broad range of gastrointestinal as well as extra-intestinal illnesses. Newborns that are tested and found to have *Escherichia coli* infection are frequently colonized by the bacteria within several hours after they are born. Most of the time, *E. coli* and also its host response may cohabit in healthy conditions and benefit from each other mutually for decades at a time. People who have weak immune systems or in instances where normal gastrointestinal hurdles are damaged, like when someone has peritonitis, are more likely to get sick from all of these *E. coli* varieties that live in their environment. The mammalian colon has a mucous layer that is home to pathogenic *E. coli*, which thrives in that environment. It is considered to be a very successful competitor found in congested locations, as it contains microflora which is found in the intestines and an anaerobe.

A significant amount of research exists on the biological makeup of this organism, but little is known about the processes by which *E. coli* maintains this favorable coexistence inside the intestines. Some researchers believe this *E. coli* may be able to take advantage of its capacity to consume gluconate which is found in the intestines more effectively than some other permanent species found in the colon. This enables it to colonize a very particular biochemical position in the human colon. It is worth noting, however, that there are numerous extremely adaptable *E. coli* strains that already have gained specialized severity characteristics, which gives it a greater potential to colonize ecological niches as well as to trigger a wide range of diseases. In many cases, these severity characteristics are carried on genetic variations that may be moved into multiple varieties to produce new mixtures of virulence determinants, or even on genetic variations that may have been moving around in the past and have now developed toward becoming part of the genome (Rendueles et al., 2011).

Diarrheogenic*E coli* species are found all over the globe, and they are very contagious. The most common mode of transmission is fecal matter, which occurs mostly after water and food have been contaminated. A pathogen called STEC, namely *E coli* O157:H7, is transmitted mostly through feces of ruminant animals such as cows, sheep, even deer, among other animals. The occurrence of human infection may be caught by the ingestion of food that is contaminated or drink, or through coming into contact with someone who has the STEC variant. Grounded beef, contact with livestock in public locations and a good example arethe petting zoos, polluted apple juice, and polluted water in leisure facilities have all been connected to epidemics. Usually, Strains of *E. coli* need a cell viability ranging from 10 hours until six days. On the other hand, the STEC variant usually has an incubation time of between three to four days. The vaginal canal of the expectant mother is frequently the entry point for *E coli* as well as some negative bacteria in newborn illnesses. It is possible to acquire gram-negative microbes in a medical setting usually via individual transmission of infection from neonatal employees or typical environmental locations. There is a wide range in the time taken by the incubation process, which may begin as soon as a baby is born up too many weeks later.

7.6.1 Preventive Measures

Proper hand cleanliness and physical segregation of infected individuals with *E coli*, particularly those who have ESBL illnesses, are recommended as standard precautions. *E coli* O157:H7 outbreaks may be prevented by properly cooking ground beef as well as avoiding unpasteurized milk. It is recommended that patients who have diarrhea consequences of *E coli*O157:H7 refrain from using leisure facilities which include swimming pools as well as waterparks for at least two weeks following their symptoms being resolved. Health officials must be alerted if O157:H7 infection occurs, particularly at child care facilities, due to the fact that this disease is worth reporting (Qiao et al., 2018). Doctors must encourage people to keep an eye on their refrigerators regarding banned items and also not to prepare the foodstuffs if they discover anything like this in the case of an *E. coli* epidemic being mentioned in the news.

Additionally, they are expected to practice safety of food by avoiding eating beef and other poultry products that are uncooked; promptly washing their hands, the working surfaces found in their kitchen, and utensils running water and adequate soap to kill germs that might have been acquired from uncooked beef during the preparation process and avoiding cross-contamination of other items in their freezers. Those who believe that may have fallen unwell from consuming recalled food items should be advised to seek medical attention immediately since signs often appear 2–7 days following consumption of the food. Those traveling should be urged to consume only drinks from canned sources and to avoid using ice because it can be from contaminated sources, as well as consuming raw food and fruits that have already been peeled in order to avoid contracting the disease. Fruits that kids have cleaned and peeled by themselves, on the other hand, are permitted. It is recommended that only mineral water be used when one is brushing their teeth.

However, if the diarrhea is persistent or is linked to stool seen to contain blood and fever which can also be absent, antibiotic medication may be advised as a preventative measure. It is difficult to utilize vaccinations to safeguard against this illness since there are so many different infections that might make people sick. There are numerous studies have shown that the vaccine that is taken orally was observed to kill the whole-cell vaccine administered in combination with both the nontoxic B component of *E. coli* gives protection against these infections in people who like traveling. The particular vaccine was authorized to be used as a traveler's diarrhea vaccine in late 2006 found in the United States, and it has been in use since then (Azevedo et al., 2014). Conservative estimates which take into account the prevalence of ETEC disease around the globe as well as vaccination effectiveness revealed that it could only help reduce about 7 percent or fewer instances of traveler's diarrhea if indeed the vaccine is used.

Key genetic components have been shown to be crucial in the development of pathogenic *E. coli*, which has led to various pathotypes susceptible of invading the digestive system, urinary bladder, and the peripheral nerves. A "base model" of *E. coli* K-12 could be used to adjust the model structure to make adjustments to particular environments but instead, allow these adapted variants to be the cause of disease in an immunocompromised person or nonhuman host by adding or subtracting various features. Genes that can be reprogrammed have made it difficult to classify *E. coli* strains into clearly defined pathotype groups.

As a consequence of a continuing evolutionary process that really has produced a highly adaptable species which are capable of invading, proliferating and destroying a wide range of surroundings. Disease-causing pathogens of *E. coli* alter host tissue processes that range from protein biosynthesis to division of cells to transcription to death, according to the study published in the New England Journal of Medicine earlier this year. Diverse poisons, signaling pathways, and cell surface features have been used as research technology to effectively understand the underlying eukaryotic processes that are influenced by various factors. *E. coli* was formerly seen as a definitive host of the intestines, but our knowledge of the processes through which this organism may cause illness has radically altered our perception of this organism.

7.7 GIARDIASIS

Symptoms of the giardia infection can include stomach aches accompanied with bloating, vomiting, as well as diarrhea. This is caused by an infection that may be found all over the globe, but it is more common where there is a lack of hygiene and unclean water. Throughout the U.S, this illness is a prevalent source of microbial contamination. Public sources of water, recreational swimming pools, and boreholes are all places where viruses and parasites may be discovered. Transmission of Giardia may be passed from one individual to the next, or from foodstuffs to person. Diseases caused by Giardia pathogens normally go away after a few days. Even after the parasite has been eliminated from the body one is still likely to experience gastrointestinal issues and complications (Piggott & Marsh, 2004). Several giardia parasite-killing medications are typically successful, although not everyone reacts. Your greatest line of defense is to avoid getting into trouble in the first place. Even if they don't show any signs, persons with this infection could still transfer the parasite to other people via their excrement. Common symptoms and signs often begin to develop after approximately one to three weeks of exposure for individuals who do get ill. Giardia infection clinical signs usually persist between two to six months at least, but might remain longer or reoccur in some individuals.

7.7.1 Causes

There are Giardia pathogens that may be found in the bowels of humans including other animals. A tough shell termed a cyst is formed around the tiny organisms before they're even eliminated in feces. This permits them to remain outside of the intestines lasting for months. Cysts break down within the victim, releasing the pathogens. Swallowing the parasitic cysts results in illness. This may be caused by drinking contaminated water, consuming tainted food, or just by direct contact with a person who is infected. Ingesting polluted water is by far the most prevalent method of acquiring the giardia infection. Throughout the globe, Giardia worms may be found in a broad range of water sources, including bodies of water bodies. Infectious giardia may enter soil and water by soil runoff, waste dumping, or animal manure. Pools or spas may get contaminated by persons who are in diapers or who have diarrhea.

It is possible to get the Giardia parasite by eating tainted food, either via the negligence of food workers who are already infected or by the consumption of raw produce that has come into contact with polluted water.

Eating is a less frequent source of contamination than water since cooking helps eliminate giardia. This is particularly true in developed nations (Pressman et al., 2020).

Individual contact – Parents while changing their child's diaper are particularly vulnerable to contracting giardia if their hands get contaminated with feces. Workers at child care facilities, where epidemics are becoming more widespread, are at risk, as are the children they care for. Anal intercourse may potentially transfer the giardia virus.

In developed nations, giardia infection is nearly never deadly. There may be long-term side effects and even death, particularly in young children and newborns. There are a few potential issues such as dehydration where the body becomes dehydrated if it may not have enough water thatis vital in performing its usual activities, which is often the outcome of really debilitating diarrhea. Not being able to flourish. Giardia infection may cause malnutrition that may impede children's healthy development if they suffer from chronic diarrhea.

Irritability to dairy: The intolerance to lactose that many persons with giardia infection suffer is the difficulty to metabolize milk sugar. Just after the illness is gone, the condition may not go away.

7.8 LISTERIOSIS

Among the most dangerous and debilitating foodborne illnesses is listeriosis. *Listeria monocytogenes* is the bacterium that is responsible for its development. According to the nations and areas of the globe, the annual incidence ranges from 0.1 to 10 cases per million individuals. Listeriosis is just a rare infection, however, the high mortality rate it causes tends to make it a serious public health issue. At winter temperatures as those found in freezers, germs that because foodborne illnesses may thrive and grow. *L. monocytogenes* can be transmitted mostly by the ingestion of contaminated food which has elevated amounts of the pathogen (Pollard et al., 2016). Human-to-human transmission of infection is also possible, most prominently from expectant women to their unborn children. Soil, groundwater, and the digestive systems of animals are all places where the bacteria may be found. Soil contamination including the use of dung as fertilizer could affect the quality of vegetables. The germs in the processed foodstuffs may also be infected during preparation, and they can proliferate to harmful levels throughout transportation and storage.

Infections with the bacterium *L. monocytogenes*, whichcauses listeriosis, are common in all nations. Listeriosis comes in two flavors: non-invasive as well as invasive. There are two types of noninvasive listeriosis which are also known as febrile listerial gastroenteritis, those that are not invasive and those that are. Diarrhea, high body temperature, migraine, and myalgia which is muscle pain are all indicators of gastroenteritis. The time required for incubation is brief. Foods having high levels of *L. monocytogenes* have been linked to epidemics of this illness. Several high-risk sections that are part of the population are susceptible to invasive listeriosis, which is a more extreme case of the illness. Expectant women, people with cancer, HIV patients, organ transplant recipients, senior members of the community, and babies are just a few examples. In this kind of sickness, there is a high suicide rate of about 20 to 30 percent. Fever, septicemia, and meningitis are among the most common symptoms. It may take anything from a few days to up to three months for an organism to become viable.

7.8.1 Causes

Common signs and the presence of the organisms in smears acquired from blood, the CSF which is the cerebrospinal fluid, infant meconium, feces, foods and livestock feed are used to make the first diagnosis of this foodborne illness. Polymerase chain reaction or the PCR test is one of many ways for detecting listeriosis in people. Listeria can only be diagnosed in pregnant women by doing blood as well as placenta investigations. Pregnant women have a 20-fold greater risk of developing listeriosis than the general population. It may lead to miscarriage or even death in the womb. Premature birth, septicemia, and meningitis could also be present in newborns. HIV/AIDS patients are now 300 times more probable to get unwell than individuals with a healthy immune system (Pappalardo et al., 2017). This is difficult to pinpoint the meal that was the real cause of the illness because of the extended incubation time. If discovered early, listeriosis may be cured. Meningitis and other life-threatening illnesses may be treated with antibiotics. Preventing infection in the fetus or infant is possible with the timely widespread use of antibiotics throughout pregnancy.

Listeria may be found in a broad variety of animals, including domestic and wild ones. Bacteria may infect raw meat, unpasteurized milk, including unwashed fruits and fresh vegetables. Listeriosis may be acquired by others who are indanger by consuming food that has been infected with Listeria bacteria. Infected food may be consumed by pregnant women and passed on to their unborn children, resulting in listeriosis inside the baby. Fresh milk, cottage cheese, pre-made meals, unwashed fresh vegetables, refrigerated diced chicken, seedless watermelon, and also before the fruit and fresh salads are linked to infectious disease outbreaks.

CHAPTER 8

PUBLIC HEALTH PRIORITY TOWARDS FOOD SAFETY

CONTENTS

First and foremost, food safety is a public health priority forhuman well-being. As a key element of the country's integrated approach to strengthening food safety, it requires an effective and efficient legal framework to ensure compliance with the law and its effective enforcement. The law also needs to be aligned with international standards and agreements. For example, countries have committed themselves to meet international food safety standards, set by the Codex Alimentarius Commission (Codex), which is jointly administered by the Food and Agriculture Organization of the United Nations (FAO) and the World Health Organization (WHO). It is well known that food poisoning is a big issue, affecting millions of people every year. Our public health care systems have done little to address this issue, as they spend over 70% of their total budget on non-communicable diseases. We need a public health policy that cuts down on food-related deaths and illnesses.

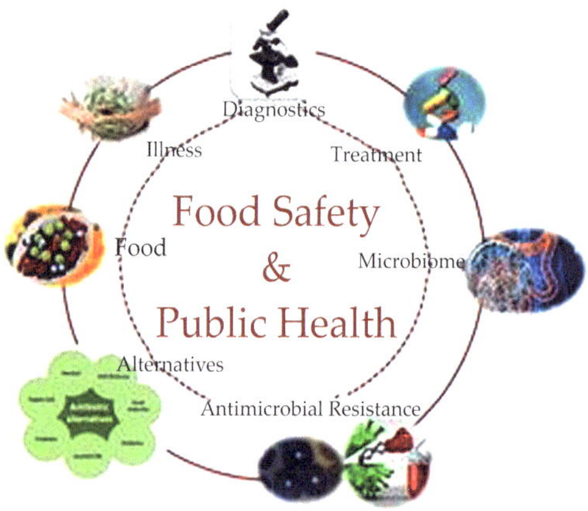

Figure 8.1: An illustration of the concerns about food safety in public health.

Source: Image by the Standard

The food supply is under constant attack from pathogens, and food safety remains a public health priority. The Food Safety and Inspection Service (FSIS), an agency of the United States Department of Agriculture (USDA), uses risk-based approaches to assess the nature and reach of problems in each establishment and impose corrective measures when necessary. However, a coordinated approach to the evaluation of microbiological testing data across all establishments producing similar products can provide even greater opportunities for population-level intervention (Antle, 2001). It is simply unhealthy to eat food that has been prepared under unsanitary conditions, or to eat stale or improperly preserved food. It is the government's job to protect the public from known dangers—and in this case, that means making sure that food purveyors are taking all the necessary precautions to ensure the safety of their products.

Indeed, food safety is a public health priority, it's been bringing the world together in one global community that's helping to create a healthier more sustainable planet. It has been not just rallying farmers and consumers around the world but chefs and kids alike. It is a longstanding practice of public health governance to monitor and enroll food production in trade and exchange. This chapter introduces the history of this practice in order to trace how it has been both persistent and changing over time. Persistent in that government presence has been continuous from the implementation of inspectors at colonial ports in countries such as Australia to recent border control measures, with contemporary public policy for food safety exercising a global reach to include nations beyond national borders and apply more optimistic attention towards science and technology. Yet, if we look closely at the historical record, the governmental approach has not always been present, nor has it always been enabling or positive.

Figure 8.2: General pictorial illustration of a variety of foodborne illnesses in the public domain.

Source: Image by Pinterest

For instance, Food-related illness, often called foodborne illness, is a common, costly public health problem. In fact, it's the leading type of global disease outbreak. It is estimated that approximately 5% of all cases of diarrhea are "foodborne" and probably represent the tip of an iceberg. Foodborne illnesses particularly impact developing countries where food can be contaminated at each step in the chain, from farm to table. Foodborne illnesses cost the world more than US$100 billion annually according to the World Bank. If we don't take action now, they will put increasing numbers of people at risk and may cause irreversible damage to some of the world's most vital ecosystems. Yet indeed, Food safety is the last public concern of the food industry, but it should be the first (Nesbitt et al., 2014).

Despite the significant progress made in food safety over the last few decades, more than 600 million people still fall ill and 420,000 people die each year from consuming contaminated food. The high burden of foodborne diseases is particularly concerning for low- and middle-income countries, where most deaths occur. We look at ways to improve food safety awareness and put in safeguards to prevent this disease - a public health priority towards food safety. Public health policy towards food safety has changed over time. The public, the professionals, and the industry increasingly agreed that food safety is important. Furthermore, recent developments in foodborne illness outbreaks have resulted in a higher degree of urgency to make sure the food supply is safe. A discussion on the changes in priority during the past century demonstrates the time-dynamic nature of public health priorities towards food safety. It also shows how this priority might further evolve based on consumers' demands and future developments in molecular biology and technology.

Tackling this issue has been identified as one of the key global challenges in food production. One solution to address this challenge is to perform a full-chain risk assessment, i.e. considering all different hazards involved along the entire food chain (Meng & Doyle, 2002). With such an approach, it is possible to identify and prevent hazards before they occur at an early stage. Increased interaction between the food chain and human societies, the intensification of agriculture, increasing world trade, globalization of the food supply and dietary changes are leading to substantial changes in the nature of foodborne diseases. Many new hazards such as *E. coli* O157:H7, Hepatitis A virus and norovirus have emerged or been recognized in recent years.

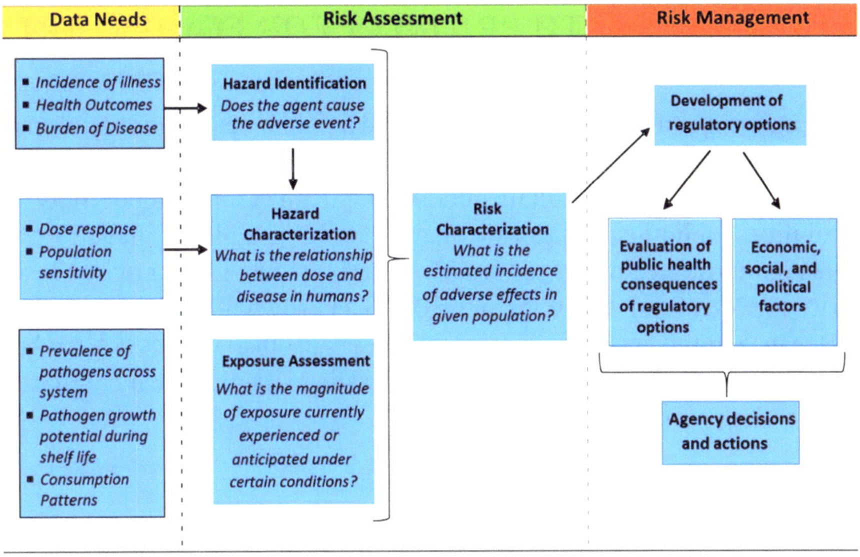

Figure 8.3: Risk assessment of foodborne diseases.

Source: Image by Springer

This chapter will majorly focus on the United States as a research basis for food safety and its public health priority measures. It is in the United States where between 60 and 81 million people get food poisoning every year. The US Centers for Disease Control and Prevention identified foodborne illnesses as a public health priority forfood safety and prevention. At its core, food safety is the cornerstone of the U.S. food supply. Maintaining a Safe Food Supply for America: A shared responsibility of industry and government, consumers have been affected by contaminated produce, melamine in milk, and salmonella in peanut butter. More specifically, the US Government has declared to take a more assertive stance by making safety a priority through regulations. Food safety is both crystal clear and a massive challenge to ensure. In addition, it is a complex subject because it has many different causes, such as from naturally occurring chemicals, infectious microorganisms like bacteria and viruses, or poisonous chemicals that can be created accidentally or intentionally (Baluka & Rumbeiha, 2016).

8.1 PUBLIC HEALTH PRIORITY FOR FOOD SAFETY IN THE CHANGING SOCIETY

In the beginning, people would not have had a public health priority towards food safety since it was not even a concept. In the future, we may have higher standards of food safety. With the increased globalization of food safety and traceability, public health priorities have also been a global matter. For as long as humans have organized into societies, we have had a public health priority towards food safety. As the world has grown and adapted, so have we. Because of this, people are safer when they eat than they ever have been before. With continuous updating of food safety practices, we continue to evolve with the times.

Figure 8.4: Historic art of food safety in the golden days.

Source: Image by IFT.org

Food is such a basic necessity that many of us think little of the convenience we enjoy. What if you had to spend three hours each day searching for food? This was the case with early man. Humans began realizing the importance of agricultural farming around 10,000 B.C. and began the domestication of animals about 8,700 B.C. The use of agricultural farming meant manpower could be used for other activities, not just finding food. This discovery has led us to our present-day techniques of agricultural production and marketing.

The importance of food safety has been known throughout history, with the Chinese protecting their rice supplies as early as 1,600 bce. It wasn't until the mid-19th century that countries began to introduce legislation and systems to ensure food safety alongside population growth and increasing demand (Mensah&Julien, 2011). Public health priority has shifted towards food safety because of new technology to identify pathogens and knowledge of their effects on the public.

The history of public health has primarily been focused on topics of mass prevention rather than individual treatment. Public health in the United States has gone through several stages of development, largely influenced by social conditions throughout the nation's history. The following is an outline of the history of public health. The practice of public health's priority toward food safety began in the 1800s when doctors and government officials recognized the connection between poor sanitation and disease outbreaks. There were many controversies over what made people sick and how to prevent illness, particularly from contamination of drinking water and methods of waste disposal.

Figure 8.5: Modern evolution of food safety management in the public domain has led to the application of technology and numerous technological aspects.

Source: Image by Food Safety Magazine

Numerous diseases such as typhoid fever, cholera, and tuberculosis were consistently spreading throughout the urban New York population causing fear and panic among city officials because it was believed these diseases were caused by bad air. By 1866, a total of 569 residents died due to Typhoid fever in Brooklyn New York. To combat this problem, the New York City Health Department was established. The Health Department was created to "promote the health of the community by enforcing laws on sanitation, infectious diseases and morbidity". The founding of health departments throughout the United States focused mainly on poor sanitation practices as a cause for outbreaks. However, many officials thought that they were spending too much energy trying to fight bacteria that they couldn't see, so they tried instead to focus on food that makes people sick. Unfortunately, tons of dairy products were not inspected by government health departments during this time.

The world has advanced, and with it, the need for food safety has become a matter of health policy. The history of food safety and public health is a long one, from the first known records of food concerns to epidemics in the US to the present day. From the history of food safety, you can see that it has truly been on the global agenda to improve health standards and keep people safe. It is easy to forget that before the introduction of public health regulations and food safety programs, many people suffered from preventable food-borne diseases. From the early Greeks to now, there has been a shift in the way many cultures view food safety. Back then, no one was worried about the germ theory of disease - people didn't know those things existed! Today we have agencies such as the Food and Drug Administration (FDA) that are responsible for regulating all food products sold in this country (McLaughlin et al., 1999).

Prior to the 1800s, food safety was a major challenge in cities around the world. Cholera, a disease spread from eating rotten food or drinking water infected with bacteria, killed thousands of people around the world every year. It wasn't until Dr. John Snow invented the water pump handle that fatal cholera outbreaks started to decline. The invention of refrigeration and pasteurization also helped keep diseases like cholera at bay in the 1900s, making it safer for people to eat meat, fish, and drink milk." Do you know how much your food is really costing you? You may not be thinking about the true cost of food, and how it affects not just your health but your pocketbook, but safe food is an integral part of a healthy lifestyle. With the advent of modern agricultural methods, many farmers now use chemicals that were unavailable even 50 years ago. These chemicals have been shown to pollute local water supplies and can affect human health, as well as causing widespread damage to the environment. We have seen what happens when we allow big companies to influence our opinions about food-related issues; this is why it's so important for everybody to understand just how much our food choices impact us in the long run. In the 19th century, early pathology work became formalized in Germany as a way to track and prevent illness. Food safety played a role in this process, and was eventually established as a more organized practice in the 1850s.

The beginning of food safety as a public health priority dates back to the Middle Ages where Bishop Thomas Gilbert recorded the diet and health of his congregation. He found that certain foods heavily impacted their health, and thus advocated for policy concerning food production, preparation, and consumption. Food safety is now a top public health priority in order to prevent major outbreaks of foodborne illness that can cause illness, death and even permanent disabilities. As global growth brings cultures closer together and builds new markets, the need for food safety grows (MacLaurin, 2001). Numerous technologies have come to the forefront in order to provide safe food to the growing world population. The issues around food safety have grown in importance over the years. The main issues that continue to persist are disease, spoilage, and adulteration. This can be seen today with the issues of tainted baby formula in China and the rise of food allergies in every country across the world.

Figure 8.6: Bishop Gilbert Thomas; the man renowned for his advent of prioritizing public health.

Source: Image by Wikipedia

8.2 PROBLEMS ASSOCIATED WITH FOOD SAFETY UNDER PUBLIC HEALTH

Over the last two decades, many countries have been faced with an increasing number of problems associated with food safety. The problem of food safety has a great impact on public health, and is of increasingly common concern to consumers, policymakers, and industry alike. In these countries there are a host of problems associated with food production, processing, storage and distribution that undermine the quality and nutritive value of food. Factors such as the inadequate disposal of human waste and garbage, unregulated and illegal use of pesticides and fertilizers by farmers, unhygienic handling and preparation practices by food handlers are just some of the issues. The result is a high incidence of water-borne diseases causing chronic symptoms, and problems with diarrhea in small children through to severe outbreaks of cholera with large numbers killed by dehydration.

Hazardous food and health risks are often perceived as being a problem associated with the poor. However, this is not always the case as the wealthy are also prone to diseases spread by food. They can also die like their counterparts in the slums if food safety regulations are not adhered to." food must be managed so as not to cause food poisoning, food poisoning is a condition when food is contaminated with bacteria or other toxins, causing complaint complaints and poisoning. The problem of food safety under public health is very serious. Physical illnesses such as diarrhea and malnutrition are a big problem all over the world (Mou & Lin, 2014). Millions of people die, mostly children, due to poor sanitation and hygiene conditions worldwide. Human contamination of public food illness is a serious matter. The problem between food safety and human health should be given more attention. Contaminated food accounts for thousands of deaths and millions of illnesses each year. Bacteria are the leading cause of foodborne illness; however, undercooking and improper storage of food can also lead to food-borne illness. Protect yourself from these problems by learning how to identify, handle and store potential health risks.

Factors that lead to the epidemiology of emerging foodborne diseases in the public sector.

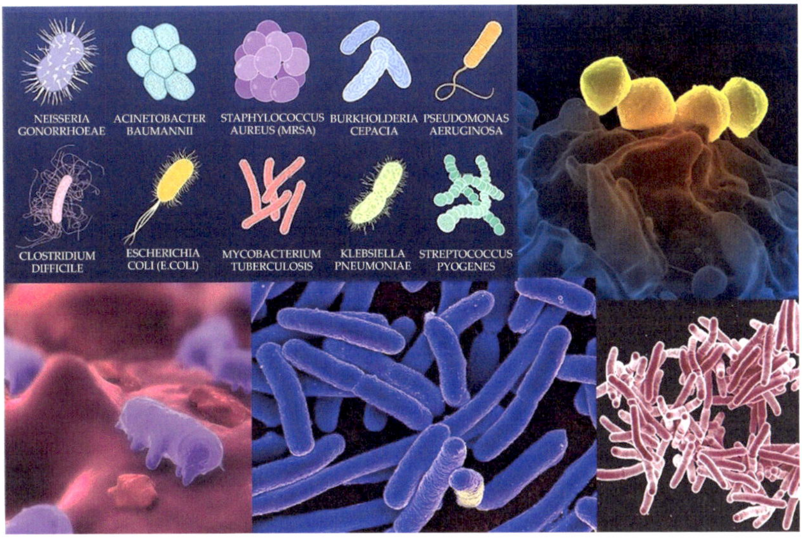

Figure 8.7: Types of pathogens. The main carries of disease causing infections.

Source: Image by p3mountains

It is increasingly recognized that foodborne illness has a major economic cost. New pathogens continue to emerge, and old pathogenic bacteria continue to evolve and remain the predominant cause of foodborne disease. Foodborne infections are of particular concern because they frequently affect large numbers of people at one time, particularly during epidemics linked to specific foods or establishments (Lytton, 2017). Foodborne diseases also have a disproportionate effect on vulnerable populations, including children, older adults and pregnant women, as a result of their developing immune systems: these groups also have a high risk of chronic complications associated with foodborne infections. Therefore it is essential to prevent foodborne illness wherever possible and we need a fundamental understanding of the factors that do lead to the epidemiology of emerging foodborne diseases in the public sector.

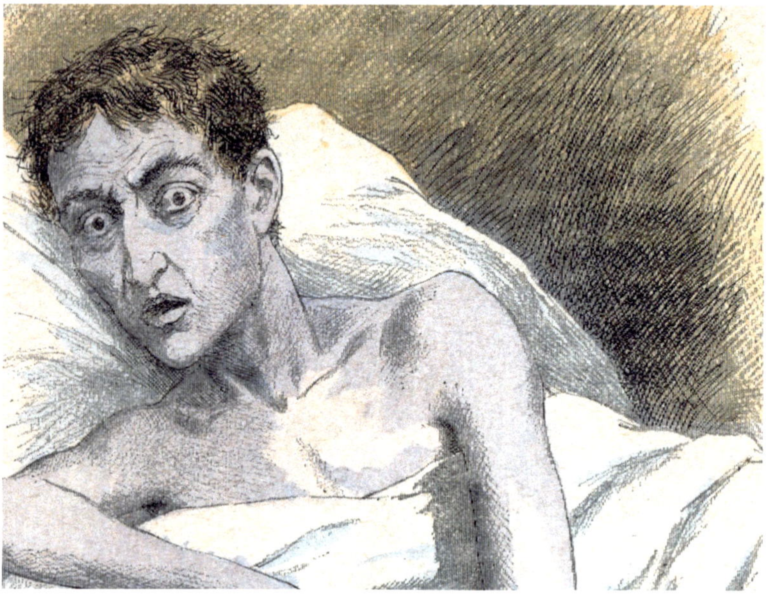

Figure 8.8: Artistic evidence of the 1832 cholera epidemic.

Source: Image by ThoughtCo

Foodborne pathogens have the potential to cause acute illness, chronic disease and death, with those most vulnerable being infants and the elderly. In recent years there has been an increase in the emergence of foodborne pathogen outbreaks. These include both new and known pathogens for which new virulent subtypes are responsible. It is critical that these diseases be discovered early so that control measures can be implemented to ensure public health. A key challenge for public health is to understand factors that lead to the epidemiology of emerging foodborne diseases. In this section, we describe current systems for surveillance of foodborne disease and identify factors that contribute to their effectiveness (Li et al., 2017). Public health system performance was measured using three criteria: timeliness, sensitivity and specificity. These measures were then compared with a third-party expert assessment of each system's ability to detect foodborne pathogen outbreaks as they occur. The panel will recommend options for improving the detection of foodborne disease outbreaks.

The emergence of new foodborne diseases (FBDs) has become an obvious problem in the public health system that needs to be addressed. In order to determine the trend of epidemiology, this study analyzed the mortality rate, number of health care facilities providing treatment, and number of physicians per hospital in different states. Results indicated that mortality rates were lower when patients were treated by health facilities that had a higher number of physicians. By expanding the surveillance strategies forfoodborne diseases and adding these strategies to existing systems, emergency responders will be able to have an early warning ofan emerging outbreak of a foodborne disease. With this strategy and procedure, the emergency personnel will be able to provide needed information to the public in order to better protect them from an outbreak of a disease.

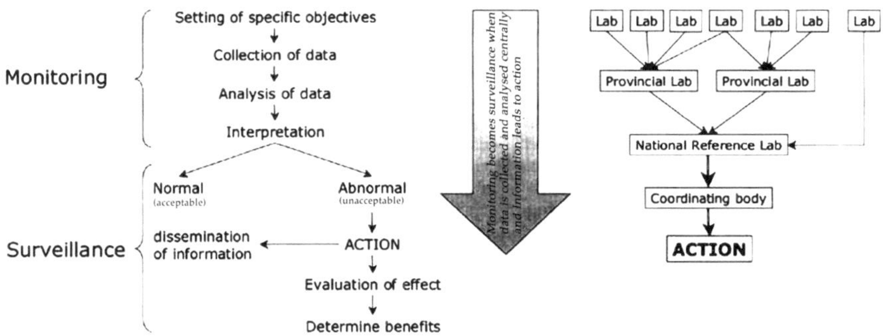

Figure 8.9: Foodborne disease surveillance measures.

Source: Image by FAO

Among other factors, the development of new methods for detecting and treating foodborne infections hasresulted in a major change in epidemiology. In addition, owing to improved surveillance, reporting and public awareness, there is a greater awareness of the risks associated with emerging infections. For example, there has been an increase in cases of norovirus infection since 1999. The full impact of this process on disease control is unclear but it is likely that there will be an increasing need for effective interventions to control outbreaks. The epidemiology of a foodborne disease is the study of the transmission cycle leading from foodborne contamination, through clinical illness to recovery. In the case of pathogenic microorganisms, two major phases are defined in this process: the pre-infection phase and the infection phase (Lam et al., 2013).

Autoimmune diseases have become a major public health issue. The prevalence of autoimmune diseases is increasing and public health strategies need to be implemented to identify new risk factors and preventive strategies.

Recently, the epidemiology of foodborne diseases has significantly changed, suggesting that we should also review other possible causes. For instance, the best place foreating out is wherepeople are able to enjoy the fresh ingredients and can also cook their food as they like it. Eating out has a series of benefits, which include nutritional value, convenience and experience. Eating at fast joints happens to be cheaper than cooking at home. When a person cooks at home, they have to incur extra costs of buying the ingredients and the time taken in the preparation of the meal.

Some of the factors include;

8.3 MICROBIAL ADAPTATION THROUGH NATURAL SELECTION

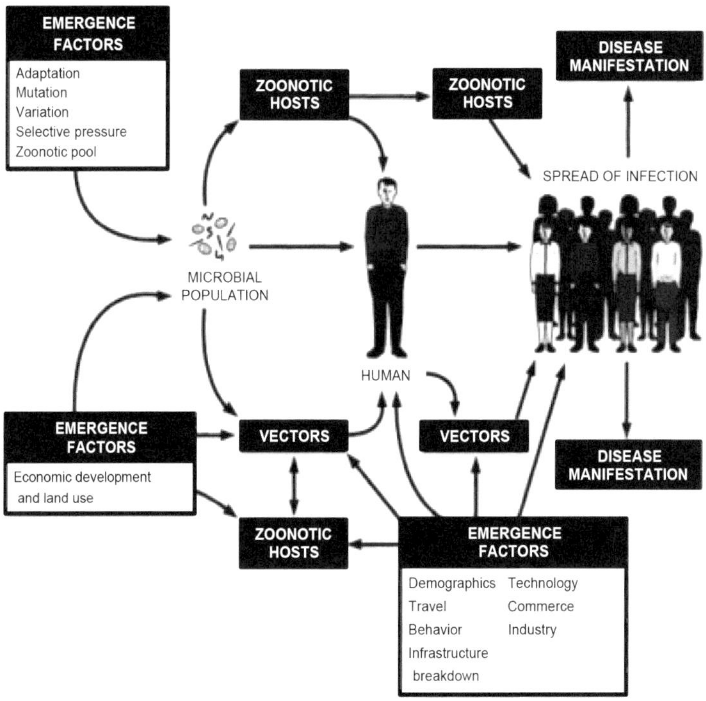

Figure 8.10: Impact of microbes on society.

Source: Image by the National Academic Press

Microbial adaptation through natural selection is one of the factors that lead to the epidemiology of emerging foodborne diseases in the public sector.

It is important that we must take preventive measures againstsuchdiseases as our food supply changes over time. Public health officials, as well as medical practitioners, need a better understanding of how certain foodborne microbes change. Recent outbreaks of foodborne diseases have been attributed to the ability of the associated microbes to adapt and adapt successfully to their environment through natural selection (King et al., 2017). The process leads to the epidemiology of emerging foodborne diseases in the public sector and should therefore be treated with the seriousness it deserves by scientists, public health officials, food industries and other stakeholders worldwide. The development of strains of foodborne pathogens with decreased susceptibility to antimicrobial agents through selective pressure may cause foodborne disease epidemiology in the public sector.

Bacteria have a gut environment that is harsh and competitive. Bacteria have diverse molecular mechanisms of antibiotic drugs resistance, thus, the ability to adaptively evolve in response to such environmental pressures due to temperature and pH changes. In recent years, the frequency of foodborne disease outbreaks in the United States had increased and caused many issues in the public sector. In the current era, there have been several advances in food safety technologies and improvements in hazard identification, risk assessment and control. However, with emerging microbial pathogen hazards, improved food safety measures do not seem to address the adaptation of microbes through natural selection to survive.

8.4 POVERTY AND POLLUTION I.E. THE LACK OF SAFE FOOD PREPARATION FACILITIES

The lack of safe food preparation facilities is a factor that leads to the epidemiology of emerging foodborne diseases in the public sector. The role of the food industry in developing nations is a significant one. The epidemiology of emerging foodborne diseases in the public sector as a result of poor sanitation and sanitation practices (Jin et al., 2020). As the food industry grows, so too does the rate at which emerging foodborne illnesses appear in the public sector. The lack of sanitary fixtures causes pollution and thus increases the rate of transmission of foodborne diseases. Sanitation is important because of the risks involved in using contaminated water for eating and drinking.

Figure 8.11: Poverty is one of the leading causes of food-related problems.

Source: Image by the Guardian

Most foods tend to be a great source of energy because the food contains healthy fats and carbohydrates. Most people prefer to eat street food as it is cheap and relatively easy to prepare. However, street foods are usually not well prepared, which may cause severe health problems. The level of training provided to vendors who offer street food isvery minimal, which often results in poor handling of the food by vendors. Also, there is no authority set up by the government that regulates the use of technology or chemicals used in preparing foodstuffs. Yet indeed, why do we exist if we are so poor, I ask myself. I am poor and I need to survive. I have to feed my family (Jespersen & Huffman, 2014). But there is no security or food in my home. So I practice scavenging as a way of living, although it's not the safest option for me and my family! We live around a dump site, you see! It's the best place to get some food while trying not to get sick due to the pollutants in the air! Researchers say none of us deserve this. We need help.

8.5 DIETARY HABITS I.E. DIETARY PRACTICES FOR RAW OR HAZARDOUS FOODS

Dietary habits and practices in the public sector have negatively affected the dietary habits and practices of the public. Foodborne diseases are spread by eating or drinking a food or beverage that is contaminated with harmful bacteria, viruses, toxins or parasites. The risk of emerging foodborne pathogens in the food industry increases from the use and misuse of industrial chemicals, as well as from microbial contaminants, especially those resistant to multiple antimicrobials. The main causes of foodborne disease outbreaks are inadequate cleaning, poor hand washing and undercooked foods. Most foodborne illnesses can be prevented by ensuring that food is properly stored, prepared and cooked.

Figure 8.12: Our dietary habits of how we handle food are of greater importance in food preservation and the prevention of diseases.

Source: Image by the Open University

The true extent of the contribution of diets toward EFDs is not known, but a number of factors appear to be increasing their importance. The food industry strives to provide value for money and is promoting convenience. This has led in part to the decline in the numbers of people preparing home-cooked meals from scratch, and thus to an increase in the consumption of ready-prepared foods. However, despite increased public awareness about food safety risks during preparation, there is evidence that this same population does not follow recommended safe-handling practices for such foods (Jacxsens et al., 2010).

8.6 DEVELOPMENT I.E. THE INTRODUCTION OF NEW FOODS THROUGH LONGER AND MORE COMPLEX FOOD CHAINS HEREBY INCREASING THE OPPORTUNITY FOR CONTAMINATION

As food supplies become globalized, the raw materials for food production and the food products themselves may be produced, processed and distributed throughout the world. This has benefits in terms of greater food availability and affordability but it also means that contamination at one point anywhere in the supply chain will affect finished products being consumed by consumers worldwide with serious consequences for public health. Development i.e. the introduction of new foods through longer and more complex food chains hereby increasing the opportunity for contamination as a factor that do lead to the epidemiology of emerging foodborne diseases in the public sector. The process of development, as focused on by this chapter includes improved water and sanitation, agricultural development and economic improvement among others. For example, through modernization efforts in agriculture, people are no more confined to single-layer subsistence agricultural systems but instead have gone industrial where they can earn good incomes through farming which leads to access to better diets and a way of life.

8.7 BIOLOGICAL AND CHEMICAL AGENTS RELATED TO FOODBORNE DISEASES

There are biological and chemical agents related to foodborne diseases, including bacteria, viruses, parasites, and protozoa from contaminated food or water; toxins from plants, animals, or microbes; and hazardous chemicals from food processing. The nature of these agents differs radically; thus, their modes of entry into foods, their means of spread within foods from foods to humans, and their methods of prevention differ significantly. An in-depth understanding of food preservation is required for successful control measures against microbiological agents such as bacteria and viruses (Jackson, 2009). Foodborne diseases remain a leading cause of illness. The nature and severity of the symptoms depend on such factors as the dose ingested and the ability of the food contaminated with a biological agent to support its growth. In this case, those who are affected by foodborne diseases have always had it wrong.

8.8 MITIGATING FOODBORNE DISEASES THROUGH PUBLIC HEALTH MEASURES

Foodborne diseases could be prevented by implementing public health measures at all stages of the food chain. Just as the meat poultry and seafood sectors have made substantial progress in reducing foodborne illness, the same can be done for other food groups. A Top 5 list of improvements that would make the biggest difference include: 1) Food safety training during farm to fork; 2) tracking of produce contamination; 3) more frequent testing; 4) improvements in sanitary design and practices; 5) a better understanding of how to control pathogens.

Indeed, Foodborne illness outbreaks have serious consequences, including death. Foodborne illnesses are preventable and all parties involved in the food production chain have a responsibility to produce safe food. Mitigation strategies for controlling foodborne illnesses have been identified. These include raising consumer awareness about food handling, improving food sanitation and hygiene along the production chain, and implementing strict regulations and surveillance systems at local, national and international levels. A key objective of food safety programs is to prevent foodborne diseases by providing good science-based risk analysis, risk management, and risk communication. There is no substitute for prevention. Risk analysis, which consists primarily of hazard analysis (identification and assessment) and risk assessment (risk estimation) is crucial to the development of science-based measures to manage foodborne hazards. By being able to identify the major causes of human illness - and knowing why they sometimes fail - scientists can then recommend measures that will allow our food systems to produce wholesome, safe foods.

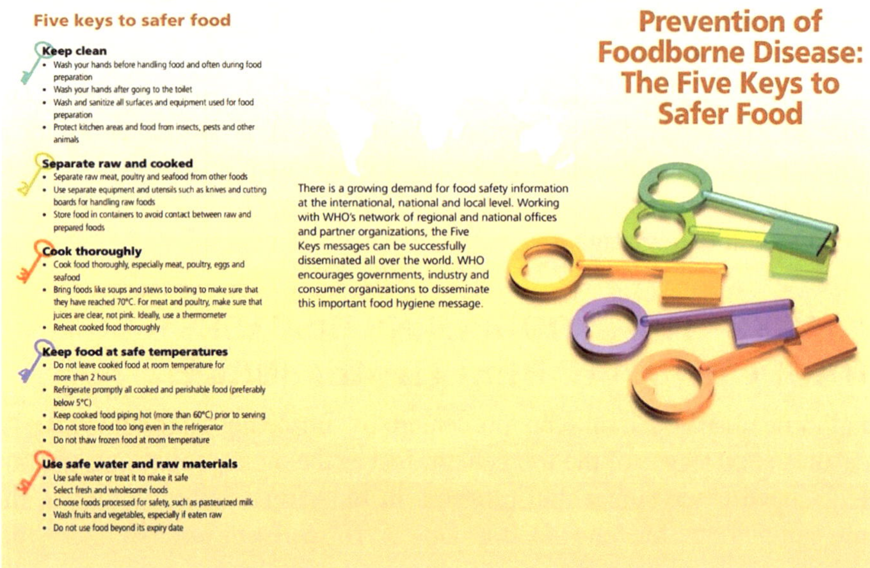

Figure 8.13: Some of the ways to attain healthier foods in the public society today.

Source: Image by WHO

Successful efforts to control foodborne diseases require concerted action and cooperation among national government agencies, food production, transportation, and processing stakeholders, professional societies, academia and public health research partners. Communication among these stakeholders is imperative to address the complex issues of food production and public health. Public health experts should also work with all of the stakeholders to develop legislation that strikes a fair balance between protecting consumers and allowing food companies to remain competitive in the marketplace. The world has changed. Food gets to our table and our stomachs more quickly and reliably than ever before. But traveling food can pick up germs along the way, putting the public's health at risk. CDC uses a One Health approach – working across human and animal health agencies – to investigate outbreaks and give real-time advice to people who need it.

According to the WHO and FAO, over 1.0 billion people (mostly younger and poorer) are affected by foodborne diseases each year. Although the number of cases is decreasing in the EU, it is increasing in developing countries. The domestic transmission of certain foodborne pathogens like hepatitis A virus, norovirus and parasite infestations continues to be higher for populations in low-resource settings. Different preventive public health measures are needed across different geographic regions as a long-term strategy to reduce the burden of foodborne diseases worldwide (Hellwig et al., 2019). The role of the public health sector in mitigating foodborne diseases has been discussed at international meetings, including the 1st and 2nd International Conference on Nutrition (ICN), the world Summit for Food and Agriculture (FAO, March/ April 2002), the World Health Assembly (WHA) and various regional meetings. More than half of the measures proposed at these meetings focus on the stages of food production, processing and preparation. Other measures are aimed at the prevention of infection in humans by increasing access to safe water and improved sanitation.

To control food poisoning hazards and minimize food pollution, it is important to practice effective food safety management. In an effort to reduce costs and increase profit margins, some companies have chosen to neglect these forms of safe management. This has led to a huge number of food accidents in recent years: for example, the human death caused by contaminated powdered milk produced by the Sanlu Group of China in 2008 and the *E. coli* O157:H7 outbreak at Chipotle Mexican Grill in 2015. In 2016, both scandalous cases were shocking reminders that if there's no guarantee of product quality and food safety, it will result in huge economic losses, not to mention its harmful effects on consumers' health. Therefore, this review summarizes how researchers backed by public health institutions have been leveraging nanotechnology to improve meat safety. The review provides an overview of recent developments on nano-scale particles (such as metallic nanoparticles, magnetic nanoparticles and carbon nanoparticles) and Zeolites that could be used in antimicrobial packaging systems (e.g., tissues embellishment with the above-mentioned types of nanoparticles; antimicrobial agents incorporated edible coatings; edible coatings incorporating inorganic layer structures; layered double hydroxides coating).

CHAPTER 9

GOOD AGRICULTURAL PRACTICES TOWARDS FOOD SAFETY

CONTENTS

Figure 9.1: Good agricultural practices ensure plants are produced in a safe manner.

Source: https://www.researchgate.net/figure/Soil-management-through-different-agricultural-practices-can-further-optimize-their-use_fig1_268747111

Good agricultural practices also known as GAPs are defined as the voluntary guidelines used by farmers during production to reduce the risk or microbial contamination that lead to food borne illnesses. All the guidelines are based on the Food and Drug Administration Guide. It is useful in minimizing microbial food safety hazards for fresh produce. There has been a notable increase in the consumption of fresh produce. The same applies to food borne illness outbreaks linked to fresh produce (Hermann et al., 2018). Most of these illnesses are linked to bacteria such as Salmonella and *E. coli*. Other than bacteria, there are parasites linked to these diseases which include Cyclospora and Cryptosporidium. This has led to buyers such as supermarkets and produce distributors requiring vendors to be audited by a third party. The certification is used to prove that the vendors have followed the Good Agricultural Practices useful in minimizing the risk of microbial contamination on food produce.

9.1 SELECTION OF PLANTING MATERIAL

Figure 9.2: The kind of planting material used in planting affects its safety.

Source: https://www.gardeningknowhow.com/garden-how-to/propagation/propgen/types-of-plant-propagation.htm

Planting good quality planting material is considered a good agricultural practice. Among the planting materials used by most farmers are seeds. Usually, farmers take a considerable amount of time deciding on the type of seeds to be used in their farms. Seeds form the background of a farm. If modern technical activities that are modernized and modified for daily farm operations do not have a constant and steady supply of high-quality seeds then they may not be able to produce safe food for consumption or they may produce safe foods but in small amounts. Seed quality is a key determinant in predicting whether the farmer will produce safe products. Seed quality can be defined as the ability of a seed to maintain high standards of vigor, purity, good generation percentage, trueness to origin and appearance. There are different kinds of seeds used by farmers and are generally grouped into seven main groups namely nuclear, breeder, foundation, registered, certified, formal and informal seeds.

Figure 9.3: Quality seeds produce healthy crops.

Source: https://www.agrotechnomarket.com/2012/04/selecting-seeds-of-good-quality.html?m=1

Nuclear seeds are the propagation material composed of mini-tubers and plantlets. *In vitro* is among the techniques used in production. These seeds are usually grown in controlled conditions. The breeder seeds are those that are produced under the direction of the main plant breeder which decides the variety hence its name. The foundation seeds are considered the descendants of the breeder seed. They are produced under conditions known to encourage genetic identification or purity. The foundation seeds are considered the ancestors of certified seeds. The certified seeds are also considered descendants of the breeder seeds. These seeds are produced under guidelines used in maintaining genetic purity and identification of the variety (Havinga & Verbruggen, 2017). For such seeds to be considered certified seeds, they have to pass certain criteria for purity given by law. Certification is done by a designated seed certification agency. Formal seeds are the modern agricultural seeds are those that have been produced by modern techniques such as genetic manipulation, treatment and hybridization. Informal seeds are those produced locally and saved by local farmers. These seeds are pure in that they do not experience any form of genetic manipulation as the method of production is purely organic. All these seeds are usually produced with the aim of ensuring that they can survive harsh environmental conditions.

Figure 9.4: Low-quality seeds are very vulnerable to diseases outbreaks.

Source: https://www.researchgate.net/figure/Poor-quality-seeds-Not-good-for-planting_fig3_342978107

Low-quality seeds are those that have poor germination percentage, low plant vigor, mixed varieties and are susceptible to attacks from pests and diseases. For this reason, low-quality seeds may not be very useful in producing safe foods for consumption as they can easily carry food borne illness agents. Growing good quality seeds and low-quality seeds together render the good quality seeds vulnerable to diseases. For this reason, the type is seeds planted will influence the farmer's success rates. However, farmers are faced with a challenge. There are many seeds both offline and online stored claiming to be purely organic and therefore need to have knowledge on how to choose seeds. Proper selection can be made when the farmer knows the method by which the seeds have been produced. By knowing the manner or the procedure followed when producing seeds, one can figure out whether the seed is hybrid, organic and heirloom. The farmer will be able to avoid selecting seeds that were produced through genetic manipulation and hybridization methods (Griesche & Baeumner, 2020).

The following steps are followed by farmers when producing the best quality seeds: The farmer removes weeds growing in the farm by hand or other methods. The weeding process involves the removal of off-type and diseased plants. During production, seed selection is usually done on the basis of important quality features including low input requirements, flavor, aroma, high yield, good yield of non-green biomass pest, storability and disease resistance. During harvesting, seeds of good quality are usually hand harvested so as to avoid mechanical damage to seeds. Hand picking is also done to avoid contamination by unwanted material and weed seeds. Once the harvesting process is completed, seeds are subjected to a threshing and a cleaning process so as to reduce impurities. Seeds are then dried up under the sun so as to remove moisture content.

Figure 9.5: Farmers can produce their own seeds.

Source: https://grain.org/en/article/6035-the-real-seeds-producers-small-scale-farmers-save-use-share-and-enhance-the-seed-diversity-of-the-crops-that-feed-africa

Seeds are then separated from mother grains during storage. There are various methods used in storing seeds. Seeds are well preserved to protect themselves from damage by insect pests. It is also done to reduce moisture content in seeds. In most cases, preservatives are added on seeds to increase their durability. Some of the preservatives include ash, eucalyptus leaves and neem. This is mostly done in local production. In the entire process, there is no room for chemical or genetic manipulation. The farmer can identify whether seeds are of good quality when having information on whether they are qualitative in nature (Gill et al., 2007). They can also be confident in seeds with known origin. The National and international seed banks is a good source of seeds. They have different types of seeds that are pure and qualitative in nature. The National government and international organization are among the bodies known to store different types of seeds from various crop varieties in seed banks. In as much as the banks store seeds for farmers on a national level, they store different types of seeds which will be difficult for them to maintain local varieties of seeds with the ability to adapt to changing local climatic conditions.

Other than the National governments and international organizations, the community seed banks and local community institutions can supply farmers with good quality seeds. There are urban and rural communities that possess seed banks used in preserving their quality seeds with some seeds extracted from indigenous plants. There are cases where small groups of people store their local seeds so as to conserve and save the seeds from a decline in quality and also to save them from extinction. There are different types of seeds leaving the farmer spoilt for choice. Farmers can also get their seeds from seed companies in the commercial sectors. These companies sell their seeds. The seed companies produce different kinds of seed types including seeds that have been genetically modified, treated or hybrid seeds.

Figure 9.6: Farmers can source seeds from various corporations.

Source: https://www.gardensall.com/garden-seeds-companies/

Though several companies produce good quality seeds, the challenge is that some companies are not reliable. For this reason, seed buyers have to cross-check and compare different labels and packets through the available seed catalogs. Certain pointers are used in helping farmers in selecting and buying seeds from commercial seed companies. Much emphasis is placed on the seed packet, seed expiration date, type of seed and the variety of seeds. The first thing farmers look at when buying seeds from companies is the kind of packaging. The seed packet usually gives information on the variety of the seed, the expiration date, the type of seed and the label. The label indicates whether the seed is GMO, hybrid, Organic or Heirloom. However, this is not safe as some online and offline stores label their seed products as organic or hybrid but in a real sense they are not (Fung et al., 2018).

Farmers are known to grow seeds that best suit the season. After properly selecting good quality seeds, the seed undergoes a selection process where they are separated according to the growing season. Seeds are usually planted on the basis of the growing season in certain climatic conditions. For instance, if the region experiences winter for longer durations than winter seeds are planted. The same applies to areas that experience longer summer seasons. However, when winter seeds are planted during summer, they may not survive or suffer from certain diseases. This is because plants have a preference for climatic requirements. There are plants that prefer cold climatic conditions while others prefer hot climatic conditions. Farmers plant seeds that do week in their native climate.

When low-quality seeds are planted, any incidents of a disease outbreak or pest outbreak the plants are left vulnerable and can easily contract diseases. They can also carry food borne illnesses and when ingested, the individual will suffer. Certified seeds have been tested and confirmed to be of good quality. Other than seeds, cutting materials are also used as planting materials.

Figure 9.7: Some plants are grown through cuttings.

Source: https://www.freepermaculture.com/plant-propagation/

Certain plants can only be planted through cuttings. For large scale farmers, they may purchase the cuttings. For small scale farmers, they have the option to purchase or make their own cuttings. When making their own cuttings, they have to first choose the plants that will be used as cuttings. Plants chosen as cuttings are often healthy. Sickly plants are not safe as they may carry food borne illnesses. Other than sickly plants, plants with lots of drooping and dry foliage are also avoided. Usually, plants chosen for cuttings tend to have plenty of new growth. Usually, good green growth is more important compared to plentiful flower buds.

Plants with several flowers are avoided when preparing cuttings as it will hinder the ability of the cutting to root itself. Plants used in making cuttings are usually large enough such that when a cutting is taken from it, the plant will not be harmed. After choosing the plant material, the next steps of preparation are very crucial as they influence the safe production of planting materials. The next step involves the preparation of containers used in holding the stem cutting for rooting. The container may be a clean pot or container. A soilless mix is often used as a planting material as it drains week and effective in providing moist conditions to facilitate rooting of cuttings (Fritsche, 2018). Ordinary soil is not used as it may contain pathogens that will kill the cutting before it takes root or the pathogen will remain in the cutting and affects the safety of food. For non-woody plants, the green stems are mostly used as cuttings. Usually, new growth is selected for cuttings as they have a faster growth rate compared to older stems.

During the extraction of cuttings, sterilized scissors and razor blades are used in making clean cuts to prevent infections. After creating their own cuttings, farmers follow certain procedures to ensure that cuttings are not contaminated. After the cuttings have been rooted, they are then transferred to the farm that has already been prepared. In most cases, the farmer will have to ensure that the land is not contaminated with any pathogens. This is done by spraying land to deal with bacteria and fungi. When purchasing cuttings, farmers often purchase from companies whose cuttings have been certified as safe and free from diseases. The choice of good quality planting material is useful in the production of safe food products.

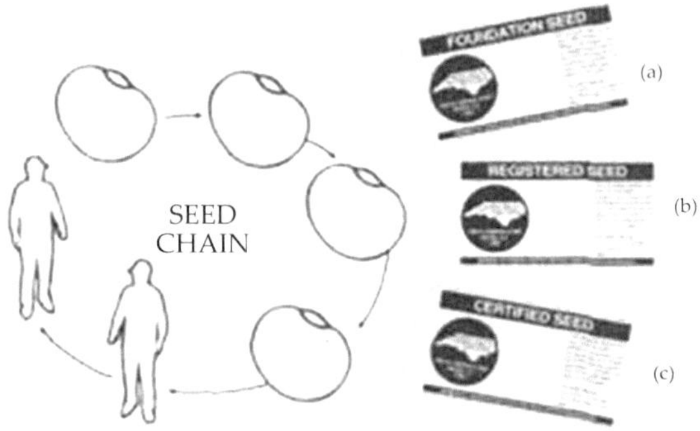

Figure 9.8: There are laws set by countries to ensure the production of quality seeds.

Source: https://content.ces.ncsu.edu/seed-and-seed-quality

In some countries, rules have been put in place to ensure the production of safe food products. For instance, the farmer needs to keep a record of the chemical used in preparing seeds and the reason for using them. The records need to indicate the treatment with dates, trade names, the active ingredient, method of application, name of operator, dosage and the reason for use. The documents should also guarantee the seeds' quality by proving that they are free from injurious viruses, diseases and pests. The farmer should also be in possession of a certificate of seed quality that states the seeds' variety purity, variety name, seed vendor and batch number. It is a requirement that the planting material be free of visible signs of diseases and pests. For in-house propagation, the scion and quality rootstock materials are to be used. It is required that propagation material to be sourced from a government-certified nursery, private nurseries, agricultural institutions and accredited tissues culture laboratories (Ergönül, 2013).

If the seeds were treated with additives or pesticides such as biocides, insecticides or fungicides then all the used products should be approved by the relevant body. It's also required that treatment of seeds be done as per the recommended technologies. When planting material is obtained from another nursery or farm then the name of the farm and date of the supply is to be recorded. It is also required that varieties known to be toxic to humans shall not be grown as they could ruin food safety.

9.2 GENETICALLY MODIFIED ORGANISMS

Over the years, it has been very difficult to produce a large amount of safe food produce. This is because the original species of plants are very vulnerable to diseases. This leads to the genetic modification of plants to make them resistant to diseases. Genetic modification of crops makes them strong so that they can survive harsh environments. However, for safety it is required that trials and planting with genetically modified crops are done if permitted by the relevant authority of the given area. Before the genetically modified crops have been successfully planted during the trials and there are no negative effects then they can full be planted in large scale farms. Farmers chosen to plant genetically modified crops with the permission of the relevant authority should get the needed documentation. It required that seed producers and farmers inform clients of the status of the products with respect to genetically modified organisms.

Figure 9.9: Genetically Engineered crops are resistant to diseases.

Source: https://www.intelligencesquaredus.org/debates/genetically-modify-food

For genetically modified crops to be considered safe for consumption, a written plan should be available for handling genetically modified materials. The plan usually sets out the strategies used in minimizing contamination hazards such as the accident mixing of adjacent non- genetically modified crops and also maintain the product's integrity. The plan is usually well laid out to enable the farmer to properly understand the instructions. Usually, the genetically modified crops are stored separately from other crops before dispatches to avoid adventitious mixing. Genetic modification of crops usually involves the suppression of certain characteristics in plants (EFSA et al., 2020). In most cases, the characteristics that are suppressed are of disease and pest resistance. If the crop is very vulnerable to certain diseases, it is modified such that it is able to resist some of the diseases. The plant is able to fight off the diseases. Genetic modification allows the production of crops that would not have survived in dry seasons to survive in such seasons. It has been useful in ensuring a continuous supply of food across the world throughout the different seasons.

9.3 APPLICATION OF FERTILIZERS AND SOIL ADDITIVES

The application of fertilizers and soil additives is covered under plant nutrient management and fertilizer use in agricultural practices. The application of good quality fertilizers in the right amount is effective in maintaining food safety. Fertilizers are usually applied according to the level of deficiency of nutrients in the soil. Fertilizers are useful in improving the health and nutrient content of plants. Inorganic and organic fertilizers are usually stored separately from harvested farm produce. Fertilizers are usually purchased from certified companies that can prove that their quality is the best. Excessive application of fertilizers on plants may lead to leaching leaving plants vulnerable to food borne illnesses. It was highlighted above that the right procedure has to be followed when applying fertilizers. There are various methods of fertilizer application that can be taken into consideration. The application methods can be placed into four main categories consisting of fertigation, placement, foliar and broadcasting.

Figure 9.10: Fertilizers influence food safety by affecting plant health.

Source: https://www.sgs.co.ke/en/agriculture-food/fertilizers/analytical-services/standard-fertilizer-tests

The broadcasting method of application involves the uniform distribution of fertilizer over a cultivated field. This method may not necessarily be effective as it will not land in soil. The foliar application method involves the application of liquid fertilizer in a spray so that it settles on the leaves of plants. There is also the placement application method that involves the application of fertilizer near plant rows in a given field. Fertigation is a unique method of fertilizer application. It involves the addition of fertilizer in the irrigation water this way fertilizer enters agricultural fields through sprinkler, furrow and drip irrigation systems. It makes use of green industrial technology. This kind of fertilizer application method enables farmers to save time, labor, money and water as they have more control over parts of land that have been fertilized and those that haven't (Diasolua et al., 2002). This method of application is very useful as it ensures the growth of plants making them very healthy reducing their chances of disease contraction. Liquid fertilizers are known to greatly reduce the amount of toxic runoff as the chemicals are usually applied with high levels of precision. This method of application is beneficial to onion crops as it is able to control nitrogen leaching while saving water, energy and nutrients through effective application of nutrients and water. With this method of application, drip irrigation is known to be highly effective as it applies water with high levels of precision to the plant when compared to furrow and sprinkler irrigation. This makes drip irrigation the best option in the introduction of fertigation to agricultural fields.

During the farmer usually makes an assessment of the biological and chemical risks related to fertilizers and soil additives used for each crop grown. Information on all assessments is usually recorded. Assessments may include samples of plants being planted and fertilizer applied. Observations are made on the effects of fertilizer plants. The information is then recorded and analyzed. The information may contain the record of significant hazards identified with such applications. Records are usually maintained for future use. If the application of fertilizers and soil additives presents significant hazards, the needed measures have to be put in place to minimize the risk of contamination to produce. The right measures are usually identified by the conditions presented. Usually, the application of soil additives and fertilizers are done on the basis of soil analysis and by the Recommendation of the National Soil Service Center or the technically component personnel or based on the farmer's experiences.

Usually, soil additives and fertilizers are selected and applied with the aim of minimizing the risk of heavy metal contamination to plants. The recommended application practices are to be followed when applying soil additives and fertilizers at the appropriate stages of crop growth. To ensure food safety, untreated organic materials should not be applied where there are significant risks of contaminating crops. When organic materials are treated on the farm, the method of application, the date and duration of treatment are noted down. There are various data noted down when using organic material. This information is useful in identifying whether food is safe for contamination. If the organic material was obtained from outside the farm then there may be a significant risk identified. For this reason, the supplier should present a document indicating that the material has been treated so as to minimize the risks of contamination of crops (Bech-Larsen & Aschemann-Witzel, 2012).

Application of untreated human sewage should not be used in the production of fresh produce. This is because it could lead to the introduction of food borne pathogens in crops. Food safety can be ensured when the facilities for mixing, loading and storage of fertilizer and soil additives and the composting of organic material to be located, constructions and maintained in a manner that ensures the minimization of the risks of contamination of sites and water sources. Maintaining records for soil additives and fertilizers with detailed information on the source, product name, date and quantity obtained, the rate of application, method of application and the name of the operator.

9.4 MANURE APPLICATION

The application of manure on agricultural lands forms part of soil fertility management. There are those who view manure application as fertilizer application though manure is very different from fertilizer in various ways. Manure contains very valuable nutrients. There is a variation in nutrients depending on the type of materials used in making the manure. Some of the nutrients found in manure include sulfur, potassium, phosphorus and nitrogen. In most cases, nutrients from manure come from the feed that the animals have eaten. Usually, a large percentage of the nutrients consumed by either plants or animals end up in manure. For instance, if waste from steers is used in making manures then more than 75% of the manure makes up ore and nitrogen while 65% is phosphorus. The amount of nutrients in manure is also determined by the manner in which manure is handled.

Figure 9.11: Contaminated manure compromises food safety.

Source: https://www.google.com/amp/s/thecounter.org/contaminated-compost-herbicide-industrial-agriculture-backyard-gardens-clopyralid/amp/

There are different kinds of manure in agriculture. Green manure is made by planting certain crops in soil after it reaches a certain stage it is cut back into soil. Manure can also be obtained from the droppings of animals or a combination of both plants and animals. Manure can be stored over time forming compost manure. Manures from different animals have different levels of nutrients. For instance, manure from poultry has high levels of nutrients while that from horses has less nutrients. It is notable that manure from beddings will have less amounts of nutrients while pure manure has large amounts. As there are different kinds of manures, there are different forms of manure namely organic and inorganic manure. There are manures made from acombination of organic and inorganic manures. Inorganic manures form part of commercial fertilizers. Organic manures are those formed from plants and animal water. For such manures, minerals are obtained from the remains of bacteria, cells and plant tissue. For such kinds of manure, the release of nutrients is slow. In soil, the nutrients are released into soil to plants as the material decomposes (Cody et al., 2012).

The selection of organic manure is very important in ensuring food safety. When manure is filled with bacteria and fungi known to cause food borne diseases, then it is a threat to the lives of human beings. Therefore, the process followed when making manure is very important. The manure can also be tested to ensure that there are no insects. A suitable manure is selected for application. There are different methods used by farmers in applying manure. The procedure used in manure application will determine food safety. The same case applies to the tools used. In most cases, a manure spreader is used as it saves time. The manure spreader has a variation in size, used depending on the amount of land to be applied. The manure spreader is usually calibrated enabling the farmer to knowthe amount of manure that can be applied. The principle of calibration used in the manure spreader is similar to the calibration of the fertilizer spreader, a planter or sprayer. This method is very simple. It involves the spreading of traps on the ground and weighing the amount of manure falling on each trap as the spreader passes over it. For farmers who have access to truck scales, they have the ability to weigh a full spreader while applying loads to the field while weighing the spreader.

Effective application of manure is done close to planting time. When application is done during this, loss of nutrients isthrough leaching, runoff or washing away is reduced. Loss of nutrients causes a reduction in the value of manure. This may potentially harm groundwater and leave plants vulnerable to bacteria and fungi among other elements. When the manure to be applied has minimal amounts of nutrients then it is usually applied in the fall before planting as such manure is not able to wash into nearby ponds and streams. Manure is kept in place while runoff is reduced during winter. Measures should be put in place to prevent the movement of organic manure into water supplies.

9.5 IRRIGATION

Irrigation is a method used by farmers when there is no continuous supply of water for growing crops. This practice is very important as it ensures the survival of crops. When irrigation is properly done, it could allow the production of safe food products. There are three broad classes of irrigation system. These include pressurized distribution, gravity flow distribution and drainage flow distribution. Under the pressurized system is the sprinkler, trickle and other similar systems in which water is conveyed to and distributed over the farmland through pressurized pipe networks. There are several individual system configurations identified by unique features such as the center-pivot sprinkler system (Chen et al., 2020). The gravity flow system works by conveying and distributing water in the field level by a free surface, overland flow regime. The surface irrigation methods have been subdivided according to configuration and operational characteristics. Although irrigation control of the drainage system through sub-irrigation is not common, it is considered an interesting concept.

Figure 9.12: Drip irrigation helps prevent the spread of plant diseases.

Source: https://www.dripworks.com/blog/how-do-irrigation-systems-work

When relatively large volumes of water are applied in soil, it percolates through the root zone and becomes a groundwater or drainage flow. When water is controlled at critical points, it makes it possible to raise the level of the groundwater to within reach of the crop roots. Individual irrigation systems have different advantages depending on their application. For the farmer to successfully conduct effective irrigation, they need to know various factors that will affect agriculture. Depending on the type of farm, the irrigation systems are designed to maximize the efficiencies and minimize capital and labor requirements. Effective land management practices depend on the type of irrigation system and its design. The manner in which land is irrigated is influenced by the use of automation, reuse of runoff, the control of captured runoff, topographical variations and field soil. Water control structures and the location of flow measurement affect water irrigation.

Healthy crops are produced when the farmer knows when to irrigate, the amount of water to be applied to soil and how to improve efficiency. Selection of the type of irrigation is affected by the location, crops being planted, topography or the land and amount of water available. This means that the compatibility affects food safety. When using an irrigation system it is important to ensure that the system does not interfere with farming operations. In most cases, farmers use small equipment or those that are powered by animals that are suitable for small fields (Bingham &Lavau, 2012). All the factors lead to the production of health crops. Economics is a factor taken into consideration when choosing the type of irrigation system. There are pressurized irrigation systems with have high capital and operating costs and utilize minimal labor and conserve large amounts of water allowing future use. High pressurized systems are mostly used to water fields with high-value cropping patterns. On the other hand, the less expensive irrigation systems tend to require large amounts of labor. Some systems cannot be used on certain types of soil or land topography. When used, it could affect plant growth. Under economics, the type of irrigation system is chosen on the basis of the cost of maintenance, expected life of the rehabilitation and annual costs including energy, maintenance, depreciation, water, land preparation, labor and taxes.

The topography of the land on its own is a factor that affects irrigation more so when surface irrigation is used. Topography of the land involves the elevation and location of water supply in relation to the field boundaries, the configurations of the field and access to various amenities. If the land has migrating herds over the land, the type of irrigation system is limited. The uniformity and slope of the land are under topographical factors. Soil moisture holding capacity also influences the type of irrigation system. Other factors include water intake rate and depth. Sandy soils are known to have high water intake rates compared to clay soils which have high moisture-storage capacities and low infiltration. Sandy soils, therefore, need frequent, smaller applications of water. Studies on soil are useful in determining the amount of water to be supplied to soil.

For instance, if soil has high moisture rates and more water is supplied then crops grown in the field then plants will be vulnerable to illnesses. Also, water supplied to plants will affect food safety. If water supplied to crops is contaminated then the plant is vulnerable to diseases. For this reason, water supplied to plants is tested to ensure that it does not contain any bacteria or fungi. This is required by law. This is to avoid toxic elements entering plants and being supplied to individuals across the world (Chandrapala et al., 2012).

9.6 CONTROL OF PESTS AND DISEASES

Under the production of safe and healthy food, the control of pests and diseases is very crucial. There are certain seasons known to exhibit outbreaks that cause the death of large numbers of plants. Trans-boundary plant diseases and pests affect food crops and destroy crops causing farmers to suffer losses and in the long run, threaten food security. Over the years, the spread of diseases and infestation of agricultural land had increased. The resilience in agricultural production systems has been pushed through climate change, trade and globalization. This is because there has been an intensification of farming activities. An outbreak of pests affects food safety as pests eat leaves of crops. As they eat, they may lay eggs on the plant. When food having eggs is eaten, then the individual eating the food is likely to suffer from food borne illnesses. The eaten leaves are also a threat to food safety through enzymes left on leaves by plants. There are various insects that attack crops including armyworms, locusts, fruit flies, cassava disease, wheat rusts and locusts.

Figure 9.13: Plant pests and diseases compromise food safety.

Source: https://www.google.com/amp/s/www.greenhousegrower.com/production/ornamentals-production/8-plant-pest-and-disease-resources-you-need-to-know-about/%3famp

There are various kinds of diseases. Some diseases are specific to certain plants while others attack all plants. Some diseases are caused by environmental conditions. For instance, when tomato plants are grown in very moist areas then they are likely to be infected. The spread of disease and pest follow three principles ways including human-migrated or trade movements, environmental forces such as weather and vector-borne pathogens. There are various methods used in controlling pests and diseases on farms. This includes the use of chemicals which involves the spraying of pesticides at a given frequency. However, there are cases where chemicals do harm rather than solve the problem. For this reason, sustainable farming methods are used in maintaining a balance between crops, pests, weeds, diseases and soil life. This is generally referred to as natural pest management and in other cases integrated pest management (Boudergue et al., 2009).

Natural pest management is useful in preventing problems pertaining to plant diseases and pests while ensuring that chemicals are kept out of our bodies. It has been useful in eliminating dependence on pesticides and chemicals for resistance. When using pesticides, farmers are advised to follow the right procedure while using the right amounts of chemicals. This is to minimize damages caused by using the chemical. There are certain factors taken into consideration when deciding on the type of chemicals to use. Plant pests and diseases can be controlled by keeping plants healthy. This is achieved by building healthy soil. Healthy soils are known to harbor friendly insects while preventing plant diseases. Planting resistant plants varieties is useful in controlling the incidence of pests and diseases. This involves the selection of seeds resistant to common diseases and pests. Proper spacing of plants goes a long way in controlling the spread of diseases and pests. When plants are placed close together, the amount of sunshine and air reaching the leaves allows both pests and diseases to thrive. When plants are planted further apart, heat gets to the soil drying it. It is also useful in controlling the spread of pests and diseases.

Timely planting has also been known to help control the spread of pests and diseases. This is because pests and diseases are known to respond to weather such as first warm day and first rains. Farmers know pest and disease patterns and are able to identify the best times to plant crops. When plants are planted earlier than usual it improves plant's ability to resist pests and diseases. Late planting leaves crops vulnerable to attacks from diseases and pests (Bouzembrak et al., 2019) (Borchers et al., 2010). There are plants that may die due to lack of food. Diseases and pests can be controlled by planting different kinds of crops or changing crop patterns. When one type of crop is planted continuously then there is a higher likelihood of disease outbreaks. Proper watering of plants can also prevent the spread of diseases and pets. Farmers are advised to water plants from below. This is because watering from above may lead to diseases living to soil splashing onto plants. Moisture on stems and leaves provides a conducive environment for the growth of diseases. Drip irrigation and flooding irrigation have been known to keep stems and plant leaves healthy.

9.7 HARVESTING AND HANDLING

Harvesting and handling of food is key to ensuring food safety. For this reason, there are special skills needed for proper harvesting, handling, grading and packaging of vegetables. The skills are useful in insuring optimum production of quality products in the marketplace. The quality of food products can be affected by the manner in which the produce was handled during harvesting, handling, packaging and storage conditions. The safety of food is greatly determined by its quality. There are various variables consumers perceive as a reflection of produce quality including freshness, taste, crispness, condition, appearance, price and nutritive value (Carvalho, 2006). For food to be safe for consumption, it has to be harvested at the right time. Timely harvesting of crops helps ensure the quality of food products is maintained. Late harvesting of plants leaves room for contamination while harvesting earlier may lead to consumption of not readily matured crops. The right tools should be used when harvesting crops. For instance, coffee berries should be picked while leaves from certain vegetables being picked.

Figure 9.14: Proper harvesting of food ensures food safety.

Source: https://foodandnutrition.org/blogs/stone-soup/harvest-front-lines/

Tools used in harvesting crops should be properly cleaned before use to avoid contamination. High levels of cleanliness should be maintained when harvesting crops. Vegetables are perishable in nature. For this reason, handling and harvesting methods are important. The quality of food is determined by the efficiency and speed of handling food on the farm. Once food leaves the farm, it is quite difficult to control food quality. Harvesting operations are mostly done by individuals skilled in it. If the farmer is not skilled then a consultant is employed to supervise the process. Pre-harvesting process involves lining up sufficient labor, preparing packaging and containers for storage, cleaning the grading or packing shade and determining whether harvesting equipment is operable. Harvesting and handling processes are very crucial as they contain biosynthetic and degradation reactions that occur during these processes. Some of the techniques used in extending post-harvest shelf life include the reduction of respiration through decreased temperatures, reduced respiration by maintaining an optimal gaseous environment, and reduced water loss by maintaining optimal relative humidity and variety selection.

All activities done after harvesting should be done with high levels of care. Workers should clean their hands and ensure the tools used during the process are clean. The same case applies when food is processed in industries.

9.8 ANIMAL MANAGEMENT

Agricultural practices also include the handling of animals. Animals produce meat, milk and eggs among other food products. With regards to meat production, high levels of care are taken to ensure that produced meat is healthy. One of the practices is ensuring that animals are healthy. Healthy animals are those that have been properly fed. This means that they have been given feeds of high nutritional value in the right rations. Giving of contaminated foods leads to animals suffering from diseases. Grazing fields are usually checked to ensure that grass is safe for consumption (Beulens et al., 2005). Rotational grazing is done to ensure controlled grazing. When one region of the field is contaminated by worms or other pests, they are taken to other sections of the field to reduce cases of contamination. The fields are then sprayed to deal with the infestation. Foods given to animals are also tested to ensure they are safe for consumption. Animals that feed on contaminated foods have to be given the right kind of treatment. Production of safe meat involves sick animals being treated. Animal management involves calves being vaccinated at the right ages to protect them from diseases.

Figure 9.15: Vaccination of animals improves their health.

Source: https://www.google.com/amp/s/grist.org/food/farm-animals-are-about-to-get-a-lot-more-shots/amp/

When the farmer identifies sick animals, depending on the type of sickness the animals will be isolated and treated. Isolation is done when there is a viral infection. When the viral infection is severe, the animal may be laid down to reduce the risk of contamination. Regular deworming of animals goes a long way in controlling diseases. Management also involves proper management of calves as they are born. Proper care is useful in ensuring that the calves survive to maturity. The safety of meat is greatly determined by handling in slaughterhouses. Meat supplied to consumers is mostly gotten from slaughterhouses that have been licensed. Slaughterhouses have to meet all the criteria set by the state for them to get a license.

Slaughterhouses have to be routinely cleaned to reduce cases of contamination. This includes routine washing with disinfectants. Slaughterhouses pick animals to be slaughtered. They look at the health and the physical appearance of the animals among other qualities. Once selected, the animals are cleaned. Slaughtering is done using the right equipment all of which have to be clean before use. Cutting tools should be sterilized (Bolek, 2020). The right procedure is to be followed when slaughtering animals. Handling meat after slaughtering is very crucial. Workers are required to be in the appropriate attire. The protective clothing limit food exposure to bacteria. Meat is cut into pieces and transported forvarious reasons and then taken to industries. The transportation process involves the use of refrigerated trucks to ensure food does not spoil during the process.

Figure 9.16: Proper management of slaughterhouses ensures the production of quality meat.

Source: https://m.dw.com/en/live-tv/s-100825

Industries then process the meat products. The process followed during the processing of meat is created by bodies governing the processing of meat. High levels of hygiene are maintained when processing meat. There are various methods of packaging used to protect meat products from exposure to bacteria.

9.9 UNDER MILK PRODUCTION

Milk production is very sensitive. Once milk is gotten from cows, all the activities after determine its safety. When milk is exposed to high ambient temperatures for long durations of time, it becomes sour. This is attributed to the presence of inherent lactic acid bacteria and microorganisms on milk vessels. With the accumulation of lactic acid, milk becomes sour and coagulates. When milk has been extracted from cows, it is raw and contains large amounts of lactic acid. It does not appears curled and coagulates when heated. Such milk is not acceptable for sale (Zanin et al., 2017). There are various factors affecting the safety of milk. Milking practices are among them whereby incomplete milking reduces milk yields and low fact contents as the last milk contains large amounts of fat. The stage of lactation also affects milk quality. If the cow suffers from mastitis, then the milk produced by it is contaminated. The composition of mastitis milk is similar to that of blood and is not safe for consumption. Animals with mastitis have to be treated before milk extracted from them is sold.

Feeding also affects milk quality. Cows that have been properly fed produce quality milk. When cows feed on weeds, there is a high likelihood that this will affect its quality. Weeds such as black jack and Mexican marigold are known to produce a bad smell in milk after consumption. Milk from cows under treatment should not be sold. This is because the antibiotics found in milk. These antibiotics may lead to allergic reactions in some individuals. Such cows should be given enough time to recover before their milk is sold. Once the milk has been extracted, its handling should be properly managed. Workers in milk factories should be dressed in the most appropriate attire. This includes clean gloves and gumboots. Containers used in milk processing should be properly cleaned and sterilized. When transporting milk over long distances, it should be kept in refrigerators to increase its shelf life. Preservatives used in milk preservation should be safe for human beings. Contaminated milk should not be sold to the masses as it could lead to illnesses. It should be disposed off (Xiang et al., 2019).

Under egg production, poultry management facilitates safe egg production. Layers should pre properly fed to maintain good health. They should also be treated when found sick. Chicks should be vaccinated to prevent illnesses. Regular dusting is done to reduce the spread of pests. A proper system is put in place when handling eggs. Eggs should be taken from chicken to prevent droppings from reaching eggs. The quality of eggs should be checked and sorted according to quality. Rotten eggs should be disposed off.

CHAPTER 10

RELATIONSHIP BETWEEN FOOD PROCESSING AND FOOD SAFETY

CONTENTS

The processing of food is different from the process of manufacturing both of which directly relate to food safety. Food processing is a method that the food will only be processed in a way, so as not to affect its taste and quality. Although numerous processing methods are generally used to produce food products with good tastes and smell on the market now, there are still a few problems that should be solved. Food safety problems threaten the health and welfare of humans in developing countries (Wilcock et al., 2004). This difference rather discusses the current condition of food safety problems, its causes and effects globally, to present an overview analysis of the relationship between food processing and food safety. It aims to minimize human mortality from food-borne diseases by giving awareness of the problem as well as discussing possible solutions for it. It also gives some recommendations for improvement of national policy implementation in order to ensure a more improved force implementation for good purposes for society.

Figure 10.1: Modern-day food processing; this stage from harvesting plays an important role in determining food safety.

Source: Image by the Statesman

Food processing is a series of operations that change, preserve or transform food in some manner. The desired result is to extend the shelf-life or to prepare it to be eaten safely without further processing. The relation between food safety and food processing can further be defined by the negativity and negative effects of unsafe foods. This relates to foodborne diseases. Foodborne disease kills more people than AIDS and malaria combined. No one should die because of a foodborne illness. Yet, each year, the World Health Organization estimates the number of foodborne illnesses is stunning. In 2010, there were 32 million cases of food poisoning in Europe. In the US, there are over 48 million cases annually. The mortality rate is approximately 30000 people per year. There has been significant progress in reducing the incidences of common foodborne illness since 1992, when the Food Safety Modernization Act was enforced (Wertheim-Heck et al., 2019). However, food safety will not be possible until every type of product is safer from farm to fork. For example carrots or potatoes can be unprocessed and immediately consumed but other products like rice require a certain amount of post-harvest processing for safe consumption.

Figure 10.2: Foodborne diseases are caused by germs and microorganisms contained in food. They are a prerogative of unsafe foods.

Source: Image by Food Safety News

Food processing involves three major operations such as (1) receiving raw foods, (2) converting raw materials into intermediate or finished form and (3) packaging the products. These operations are common to all three forms of commercial food processing. The process of food processing is generally performed by industry groups involved in the manufacture of different food products. The act combines raw food materials, such as fruits and vegetables or slaughter products, with other ingredients to create marketable food products for sale in a range of retail stores (Wahyuni et al., 2019). Food processing typically takes clean, harvested crops or slaughtered animals and uses these raw materials to manufacture edible items that can be easily prepared and served by the consumer. The processing techniques and control measures adopted by the food manufacturers to ensure a higher shelf life of their products and to ensure stable, quality products at the time of consumption are discussed in this chapter. Control of potentially hazardous microbial food micro-organisms is an essential step in food processing.

Figure 10.3: With well processed foods, we easily achieve food safety.

Source: Image by Public Health Notes

Food safety is very important to the lives of people and their life quality, so the food producers must follow some requirements and principles. Food safety signifies the hygiene and safety of food coming from the source, before any food processing activities have been initiated. The objective is to prevent food poisoning. The main sources of food poisoning are contamination from improper handling including poor personal hygiene by workers and unsafe practices in the processing of raw materials and finished products in factories. There could also be spoilage caused by microorganisms like molds etc (Tirado et al., 2010). Food safety has been a topic of great concern lately, with sprouting recalls, unsafe field-grown produce, and food poisoning a constant threat. Many of these harmful products can be avoided by understanding the relationship between food processing and food safety. Food processing adds value to food and makes it more useful for humans. However, with this process, the food may become dangerous for people's health. For example, if a certain raw material has certain pathogens which are destroyed when cooked, the product of food processing would have no pathogens and thus be safe for consumption.

Figure 10.4: Most people will just eat foods without the knowledge of how the particular food has been processed.

Source: Image by Healthline

Consumers are generally concerned about the safety of the food they eat, particularly from the perspective of infectious food-borne diseases. Although countries have made continued efforts to improve food safety, there is still a lack of understanding of a number of them and their relationships to food processing as well as a number of regulatory questionable and undefined issues. The issue is exacerbated by improper handling in the distribution system. Food processing and food safety are inseparable throughout the world, because of the high risk of food borne diseases, there are lots of regulations for food production. Food processing products must be hygiene consistently, to ensure the security and quality of goods. To summarize, there are three types: physical, chemical, biological, each type is important for safety goals in processing. In most cases, food processing is an integral component in commercial food production and the use of a variety of technologies combine to make possible the production of an enormous range of different safer foods.

Figure 10.5: Clean processing of meat and meat products helps uphold food safety.

Source: Image by Alabama Cooperative Extension System

Operations relating to food safety can range from the slaughtering and butchering of animals, growing and harvesting of plants, to the packaging and labeling of almost any edible product. All these operations result in food that is safe to consume. Food processing makes food safe by eliminating pathogens, bacteria, parasites and viruses that are known to cause illness or disease. Good food processing procedures can help reduce and eliminate the risk of foodborne pathogens in food (Todt et al., 2009). Food processors should focus on the prevention of unscheduled downtime, such as a recall due to foodborne illness, rather than concentrating on detecting problems after they develop. Lawmakers have passed laws that define the process of how Foods should be handled to avoid contaminating them with bacteria or germs. Food safety is the study of the conditions that can lead to food-borne illnesses. When you eat food, you place your health in jeopardy if it is not handled safely.

10.1 CURRENT CONDITION OF FOOD SAFETY PROBLEMS

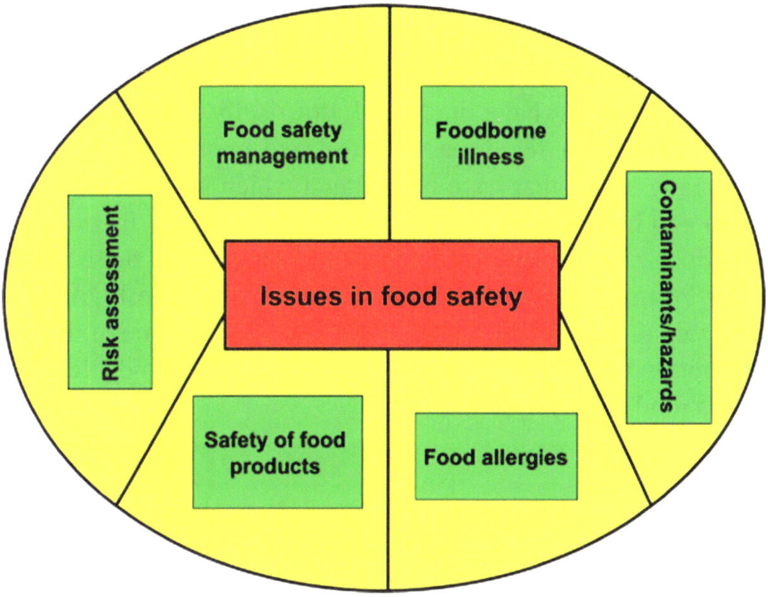

Figure 10.6: A graphical illustration of the current issues that affect food safety.

Source: Image by Science Direct

The current condition of food safety problems in relation to food processing is very alarming. For example, mishandled and unrefrigerated foods can cause food poisoning and other diseases over the course of time. According to research conducted by CDC, one in every six Americans get food poisoning every year that results in approximately 9,000 deaths (Souder). Although the government has strict regulations about how to prepare products such as meat, poultry, eggs and dairy for cooking, it is still not enough to control the rate of food poisoning. So it must be restaurants' responsibility to implement effective methods to keep those products from getting contaminated by bacteria. The possible outcomes can bring out the fact that the major issues are processed meats, eggs, leafy vegetables and sprouts. There is no denying the fact that food safety problems are growing more and more serious under current circumstances.

These problems include unsanitary equipment, insufficient knowledge about safe food processing procedures and cleaning of equipment, as well as a general lack of education within the industry. Food safety is a crucial concern for the global society and has received an increasing amount of attention as more and more recalls occur. Relevant bodies are focused on how companies were handling food safety problems at both the processing and retail levels, and how they were being regulated. In the earlier days, Food safety problems have increased drastically from 2002-2006, as reported by the Center for disease control and prevention (Sinha, 2009). There are many issues that have been found which indicate a lack of food preparation and handling procedures by not only individual consumers but the food industry in general. The CDC further reports that these issues have led to over 76 million illnesses per year in America. The FDA released a report that said Food-related incidents have grown to be one of the largest – yet underreported – results from what's considered unavoidable failures in our nation's food supply system.

Figure 10.7: Undercooked meat is one of the leading challenges to food safety.

Source: Image by Prime 13

But food safety problems aren't limited to what you eat in restaurants or get in take-out form. Recent surveys have shown that almost 40% of Americans don't cook fresh meats and vegetables at home--in part because they don't realize the risks they're taking totheir health when they consume them raw. "With so many people eating out these days, it's crucial that we work hard to make sure the nation's food supply is safe," says FDSA President and CEO Elmer Haskins. "Consumers deserve to know what they're getting when they hand money over to a waiter or stand in line at a fast food restaurant."

Moreover, there have been several events, events which have portrayed the rot in food safety. The Mad Cow Disease scare in Great Britain and a large *E. coli* outbreak in Germany in the summer of 2011 have heightened public awareness of food safety problems. There are over 700 different forms of food poisoning, with salmonella being one of the more common infections. Serious diseases such as listeria are on the rise (Associated Press, April 12, 2013). Because of globalization and frequent travel by people, communicable diseases can easily spread across national borders. This means that the U.S. Food and Drug Administration (FDA) must constantly monitor food preparation and travel for incoming visitors from other countries. Many argue that The Most Common Food Safety Problem may be a lack of knowledge. In the U.S., there are approximately 48 million cases of foodborne illness per year, and about 128,000 people are hospitalized (Sadiq & Beauchemin, 2017). The number one reason for foodborne illness is lack of personal hygiene, which includes washing your hands properly before handling food. Above all, the most important thing when it comes to preventing foodborne illness is the careful handling of food after it has been prepared. The real problem is that government agencies like the FDA are underfunded, and incapable of policing our food supply system.

Moreover, in 2008, there was a case of *E coli* that caused over 200,000 people to get sick. A year later, a tainted batch of chopped ingredients resulted in at least 16 people landing in the hospital. Then more recently, diners had an unpleasant experience with a bug found in McNuggets from KFC. As you can see, there are numerous problems with cleanliness and hygiene that have caused substantial pain for many people. Like as if to add an insult to an injury, the dangerous bacteria, Clostridium botulism, is especially prevalent in fresh-cut sliced mushrooms and has now been linked to 36 illnesses and 10 deaths. This recent outbreak of Botulism was diagnosed in California; the thirty-six individuals who have become ill all ate a garlic chicken salad with pre-packaged chopped onion and garlic in olive oil sold in supermarket salad bars that were made by Triple T. Another major food producer who uses packaged ingredients from Macro became aware of the problem when an employee saw one of the bags of chopped onion and garlic that appeared to be growing fungi. Thus far, no symptoms have been reported for these cases, however, all consumers are encouraged to immediately remove these ingredients from their homes.

According to a report by the U.S. Department of Agriculture, 97 percent of all beef, poultry and pork in America are contaminated with pathogenic substances like *E. coli* and Salmonella. Almost half of those who contract food-borne illnesses die as a result. It is now evident that the food supply must be regulated and supervised, and it is not something that can simply be fixed by the industry itself. The Agriculture industry must focus on food safety problems in relation to processing plants that further damage the already damaged food, and focus on fixing the food processing system. A majority of Today's food safety problems are due to the rapid growth of food processing and handling. According to current statistics, there are over six million cases of people getting infected by food borne illness every year, resulting in thousands of fatalities (Smith &Pillai, 2004). The status quo of today's health systems is that food processing companies are responsible for ensuring the safety of their own products while government agencies only collect statistics of reported food-borne illnesses and perform random inspections.

10.2 CAUSES AND EFFECTS OF FOOD SAFETY IN RELATIONS TO FOOD PROCESSING

Indeed, Food safety is not just about production and testing, but also it's about a process that processes concerns in the right manner and at the right moment. Food safety is a phenomenon in the society that takes care of health of the customers while they are working on their product. Food safety is a process that many food producers and manufacturers began taking more seriously in 1993 when the first major outbreak of *E. coli* occurred. The notable *E.coli* strain made people ill; this was in 1993, twenty-three years ago. The most common way food becomes unsafe to eat is due to improper storage, preparation, or handling; products that have been exposed to dangerous bacteria or contaminants can cause food-borne illnesses to those who consume them. The major concerns over the last few decades were with *E.coli* and salmonella, but as time progressed other hazards came into play such as pathogens, pesticides, and mycotoxins. It is crucial for companies to pay attention to every detail involved in their processes since incorrect preservation can lead consumers to get sick from foods containing mycotoxins.

Figure 10.8: Poor storage of foods after processing also results in unsafe foods.

Source: Image by NDTV.com

Food safety can further be broken into two categories, human and animal. Human food safety relates to how food is handled and processed both before it reaches the retail market as well as after it has been purchased. On a smaller scale, it includes things such as how food is stored, handled and prepared in the home. The safety of our food could be understood as a relatively new phenomenon and is still evolving, concepts of safe food can be traced all the way back to the beginnings of human society. In the past and even today, food sanitation has been influenced by cultural characteristics, religious practices, climate and geography. Food safety is quite a recent issue as it wasn't until around the beginning of the 20th century that methods and techniques were developed to ensure food was handled safely (Sargeant et al., 2009). Even though food safety is a largely perceived risk, it can have devastating effects on the producers, distributors and consumers of food. Food safety is a protective measure done to ensure that marketable and consumable food is served to consumers. However, hunger and poverty remain the underlying factors for making an individual vulnerable to foodborne diseases. Without adequate government or private programs, donors and international organizations play a key role in reducing malnutrition.

10.3 PROCESSING TECHNIQUES AND CONTROL MEASURES

Farmers and local processors should be trained in good agricultural practices (GAPs) and safe food-handling practices for minimizing the risk of hazards entering ready-to-eat products. Some food products e.g. processed meats and meat products remain ubiquitous throughout today's modern diet. Some countries even encourage the practice by providing necessary resources for the production of these products. For example, as early as the late 1700s, Lyon, France implemented a law mandating that sausage factories be within 5 km of the slaughterhouse to aid in the transportation of their processed meats.

Some other mechanisms can also be personally initiated. One can opt to prepare foods in their own home and reduce your loved ones of food poisoning eating out. By bringing a bit of the restaurant kitchen into your home, you can lose the fear associated with cooking with raw food products such as poultry, beef, and seafood. Much of the concern around these items is pertaining to cross-contamination. However, the USDA has implemented some simple strategies to help stop the spread of bacteria (Redmond & Griffith, 2004). Processing techniques are the procedures and operations performed on products both prior to and after harvesting. Control measures refer to the specific steps taken to ensure that raw or fresh produce is safe and suitable for consumption, there are three major groups of control measures: Physical, Chemical and Biological Control Measures. This includes the chemical freezing of foods from the industrial perspective.

The process of freezing raw materials is based on the principle of reversible freezing, which generally is possible only at temperatures lower than −40 ° C. 99% of all food products will freeze in this range without destroying the quality and without causing changes in taste and organoleptic properties. Some products can be frozen by pushing them to temperatures near −50 ° C and up to −70 ° C. However, their low technical feasibility and high electricity consumption do not make them economically attractive (except for some specialized products). In order to keep foods fresh, their temperature should be controlled between 0 and 4 ° C, respectively. This requires a constant temperature and a very efficient heat transfer between the product and the storage medium.

With a growing focus on food safety and an increased risk of costly adulterations, food processors are searching for more cost-effective methods of ensuring product quality. Consumers have become more sophisticated and demanding for safe, high quality, "clean label" or natural alternatives to synthetic ingredients. Processing techniques and control measures from the starting material to end-product are important factors in consumer perception of food and beverage products (Rendueles et al., 2011). The success of beverages and snacks depends on consumer perception regarding their health benefits. Consumers look at dozens of attributes when purchasing food, beverage and snack items. Food science can help us understand how processing techniques and control measures impact consumer perception.

10.4 REGULATIONS FOR FOOD PRODUCTION AND PROCESSING

The Food Safety Modernization Act is a bill signed by President Obama on January 4, 2011 and is the first-ever comprehensive reform of America's food safety laws. Although there are many benefits to the new system, it will also be very costly to businesses which bring up the question: How can businesses pay for their share of the cost? Another argument is that food safety has already been taken care of before this bill however there are still many food safety problems that need to be discussed and fixed. Although there are solutions to this problem, it is very controversial because some people believe we shouldn't spend the money on fixing something that should have been solved before now.

Figure 10.9: President Barack Obama signing into law the food safety modernization act bill.

Source: Image by Wikiwand

The Guide for Food Production and Processing gives food production and food processing plant personnel the tools they need to prevent foodborne illness in their facilities. FDA's goal for this guide is to help plants control hazards and assure safe food by giving them the latest information on its regulations, standards, recommended practices and other guidance. Food and Drug Administration (FDA) rules affect a wide variety of businesses. These rules range from food production and processing to the development of medical devices and drugs, cosmetic labeling, facility inspections, post-market surveillance of medical products, and the administration of food labeling and claims (Qiao et al., 2018).

10.5 THE FUTURE AND TECHNOLOGICAL ADVANCEMENTS ON FOOD PROCESSING THAT RELATE TO FOOD SAFETY

The future of food processing is here and it's based on making the production of food faster, easier, safer, and better. With new innovations in food safety alternative methods are being developed to ensure that the produce we consume is wholesome and uncontaminated. These innovations will ensure that consumers continue purchasing safe, healthy foods while food manufacturers keep up with break neck demand. Food safety is a global concern. As food travels from the farm to our homes, it may come into contact with foodborne pathogens such as listeria and *E. coli*, which can pose serious health risks. Food processing technologies are being developed to keep up with the need for better food handling practices and minimizing the risk of contamination and cross-contamination.

Figure 10.10: Some modern advancement in food processing technology helps cater for food safety.

Source: Image by Food Navigator

Food processing has been carried out successfully in recent years and much research has been conducted in this field. It provides solutions to many problems, such as reducing the energy needed, rescuing surplus or spoiled foods, or improving the texture or nutritional value of foods. Many view food safety as a grocery store issue; however, it is something everyone should be aware of. The risk for food borne illness occurs at any place in the production and distribution system. This is always an important subject to address, especially in light of globalization and the increased opportunity for contamination. It has become vital for the government and industry to work together so that people are able to enjoy healthy and safe food that can be consumed with the least amount of risk possible (Piggott & Marsh, 2004).

Less than 2 percent of the U.S. population suffers from a food allergy, yet we all want to enjoy the benefits of eating fresh fruits, vegetables and other products. To this end, there is ongoing research in the entire food industry, from pre- to post-harvest processes, to reduce the combined risks of poor sanitation and contaminated water supplies that lead to contamination at some steps in the food production process. Through technological advancements, food will be processed in ways it never has before (Pressman et al., 2020). In order for the public to feel safe in their food, there is a need for a new organization to set forth regulations and guidelines on how current technology can be used. Without proper regulation and supervision, the general public would rather not take any risks on food safety; therefore, people would eat less fresh foods, which is what they truly want.

Future innovations in food processing will help to improve food safety. Innovations such as bar codes, standardized labeling on all products, improved packaging design and automation will contribute to a safer food supply and better consumer awareness. With bio-products, new testing capabilities, and efficient packaging methods, food products are constantly improving their safety and quality. Advancing food processing methods allow for more food to be processed from scratch and served in restaurants. Food processors and end-users agree that food safety is the number-one factor limiting further growth in the industry.

Frankly enough, Food is the biggest business out there. And the food industry is enjoying increased sales and a wider reach through the benefit of new technology. Food processing has come a long way from stone-grinding corn to making a variety of products. It can be messy and risky with germs, bacteria, and viruses everywhere. The only behavioral barrier to having a variety of food is that many people find the preparation time to be too long. There are ways to reduce this time dramatically and thus having to cook as much as one desires is now possible. First, science and technology have made progress in food processing and food cooking. Technological advancement has allowed for innovations to be done on our food. We have witnessed several scientifictechnologies improve the way we deal with our food (Pollard et al., 2016). The second way this can be brought about is through processes such as microwaves, freezing, and refrigeration. These processes make it possible for us to preserve our foods for longer periods of time. These two factors combined will give us the opportunity of getting ahead with preparing our food because we can get them home quickly and cook them in small amounts of time. The third step will be when we get home we can store them in the refrigerator so they wouldn't spoil easily. We would then take out what we would want to eat and microwave it just enough so that it would be ready in minutes once heated up by the microwave.

Finally, Scientists are working hard to develop tools, such as DNA Analysis, to combat foodborne illnesses. These technological advancements further help eliminate the threat of bacteria in food products (Pappalardo et al., 2017). More so, nowadays, most farmers are using simple yet effective practices that ensure food quality to ensure that potential poisons would not be found in harvested foods. This means there is less chance for human illnesses caused by unsafe food products.

REFERENCES

1. Ambrus, A. (2016). International harmonization of food safety assessment of pesticide residues. *Journal of agricultural and food chemistry*, *64*(1), 21-29.
2. Azevedo, I., Albano, H., Silva, J., & Teixeira, P. (2014).Food safety in the domestic environment. *Food Control*, *37*, 272-276.
3. Antle, J. M. (2001). Economic analysis of food safety. *Handbook of agricultural economics*, *1*, 1083-1136.
4. Baluka, S. A., &Rumbeiha, W. K. (2016).Bisphenol A and food safety: Lessons from developed to developing countries. *Food and Chemical Toxicology*, *92*, 58-63.
5. Bech-Larsen, T., &Aschemann-Witzel, J. (2012). A macromarketing perspective on food safety regulation: The Danish ban on trans-fatty acids. *Journal of Macromarketing*, *32*(2), 208-219.
6. Bingham, N., &Lavau, S. (2012). The object of regulation: tending the tensions of food safety. *Environment and Planning A*, *44*(7), 1589-1606.
7. Boudergue, C., Burel, C., Dragacci, S., FAVROT, M. C., FREMY, J. M., Massimi, C.,&Avantaggiato, G. (2009). Review of mycotoxin-detoxifying agents used as feed additives: mode of action, efficacy and feed/food safety. *EFSA Supporting Publications*, *6*(9), 22E.
8. Bouzembrak, Y., Klüche, M., Gavai, A., & Marvin, H. J. (2019). Internet of Things in food safety: Literature review and a bibliometric analysis. *Trends in Food Science & Technology*, *94*, 54-64.

9. Borchers, A., Teuber, S. S., Keen, C. L., & Gershwin, M. E. (2010). Food safety. *Clinical reviews in allergy & immunology*, *39*(2), 95-141.
10. Bolek, S. (2020). Consumer knowledge, attitudes, and judgments about food safety: A consumer analysis. *Trends in Food Science & Technology*, *102*, 242-248.
11. Beulens, A. J., Broens, D. F., Folstar, P., &Hofstede, G. J. (2005).Food safety and transparency in food chains and networks Relationships and challenges. *Food control*, *16*(6), 481-486.
12. Carvalho, F. P. (2006). Agriculture, pesticides, food security and food safety. *Environmental science & policy*, *9*(7-8), 685-692.
13. Chandrapala, J., Oliver, C., Kentish, S., &Ashokkumar, M. (2012). Ultrasonics in food processing–Food quality assurance and food safety. *Trends in Food Science & Technology*, *26*(2), 88-98.
14. Chen, J., Sun, R., Pan, C., Sun, Y., Mai, B., & Li, Q. X. (2020). Antibiotics and food safety in aquaculture. *Journal of Agricultural and Food Chemistry*, *68*(43), 11908-11919.
15. Cody, M. M., Gravani, R., Edge, M. S., Dooher, C., & White, C. (2012). International Food Information Council Foundation food and health survey, 2006–2010, food safety: a web-enabled survey. *Food Protection Trends*, *32*(6), 309-326.
16. DiasoluaNgudi, D., Kuo, Y. H., & Lambein, F. (2002). Food safety and amino acid balance in processed cassava "cossettes". *Journal of Agricultural and Food Chemistry*, *50*(10), 3042-3049.
17. European Food Safety Authority (EFSA), Maggiore, A., Afonso, A., Barrucci, F., &Sanctis, G. D. (2020).Climate change as a driver of emerging risks for food and feed safety, plant, animal health and nutritional quality. *EFSA Supporting Publications*, *17*(6), 1881E.
18. Ergönül, B. (2013). Consumer awareness and perception to food safety: A consumer analysis. *Food control*, *32*(2), 461-471.
19. Fritsche, J. (2018). Recent developments and digital perspectives in food safety and authenticity. *Journal of Agricultural and Food Chemistry*, *66*(29), 7562-7567.
20. Fung, F., Wang, H. S., &Menon, S. (2018). Food safety in the 21st century. *Biomedical journal*, *41*(2), 88-95.
21. Gill, S., Veinot, J., Kavanagh, M., & Pulido, O. (2007). Human heart glutamate receptors—implications for toxicology, food safety, and drug discovery. *Toxicologic pathology*, *35*(3), 411-417.

22. Griesche, C., &Baeumner, A. J. (2020).Biosensors to support sustainable agriculture and food safety. *TrAC Trends in Analytical Chemistry*, *128*, 115906.
23. Havinga, T., &Verbruggen, P. (2017). Understanding complex governance relationships in food safety regulation: The RIT model as a theoretical lens. *The ANNALS of the American Academy of Political and Social Science*, *670*(1), 58-77.
24. Hermann, C. A., Duerkop, A., & Baeumner, A. J. (2018). Food safety analysis enabled through biological and synthetic materials: a critical review of current trends. *Analytical Chemistry*, *91*(1), 569-587.
25. Hellwig, M., Humpf, H. U., Hengstler, J., Mally, A., Vieths, S., & Henle, T. (2019). Quality Criteria for Studies on Dietary Glycation Compounds and Human Health: Opinion of the Senate Commission on Food Safety (SKLM) of the German Research Foundation (DFG). *Journal of agricultural and food chemistry*, *67*(41), 11307-11311.
26. Jackson, L. S. (2009). Chemical food safety issues in the United States: past, present, and future. *Journal of agricultural and food chemistry*, *57*(18), 8161-8170.
27. Jacxsens, L., Uyttendaele, M., Devlieghere, F., Rovira, J., Gomez, S. O., &Luning, P. A. (2010). Food safety performance indicators to benchmark food safety output of food safety management systems. *International Journal of Food Microbiology*, *141*, S180-S187.
28. Jespersen, L., & Huffman, R. (2014).Building food safety into the company culture: a look at Maple Leaf Foods. *Perspectives in Public Health*, *134*(4), 200-205.
29. Jin, C., Bouzembrak, Y., Zhou, J., Liang, Q., van den Bulk, L. M., Gavai, A.,& Marvin, H. J. (2020). Big Data in food safety-A review. *Current Opinion in Food Science*, *36*, 24-32.
30. King, T., Cole, M., Farber, J. M., Eisenbrand, G., Zabaras, D., Fox, E. M., & Hill, J. P. (2017). Food safety for food security: Relationship between global megatrends and developments in food safety. *Trends in Food Science & Technology*, *68*, 160-175.
31. Lam, H. M., Remais, J., Fung, M. C., Xu, L., & Sun, S. S. M. (2013). Food supply and food safety issues in China. *The Lancet*, *381*(9882), 2044-2053.

32. Li, J., Cui, N., & Liu, J. (2017). Gutter oil: an overview of Chinese food safety issues and policies. *Global health promotion*, *24*(3), 75-78.
33. Lytton, T. D. (2017). The taming of the stew: Regulatory intermediaries in food safety governance. *The ANNALS of the American Academy of Political and Social Science*, *670*(1), 78-92.
34. Mou, Y., & Lin, C. A. (2014).Communicating food safety via the social media: The role of knowledge and emotions on risk perception and prevention. *Science Communication*, *36*(5), 593-616.
35. MacLaurin, T. L. (2001). Food safety in travel and tourism. *Journal of Travel Research*, *39*(3), 332-333.
36. McLaughlin, M. J., Parker, D. R., & Clarke, J. M. (1999). Metals and micronutrients–food safety issues. *Field crops research*, *60*(1-2), 143-163.
37. Mensah, L. D., &Julien, D. (2011).Implementation of food safety management systems in the UK. *Food control*, *22*(8), 1216-1225.
38. Meng, J., & Doyle, M. P. (2002).Introduction.Microbiological food safety. *Microbes and Infection*, *4*(4), 395-397.
39. Nesbitt, A., Thomas, M. K., Marshall, B., Snedeker, K., Meleta, K., Watson, B., & Bienefeld, M. (2014).Baseline for consumer food safety knowledge and behaviour in Canada. *Food Control*, *38*, 157-173.
40. Pappalardo, A. M., Copat, C., Ferrito, V., Grasso, A., &Ferrante, M. (2017). Heavy metal content and molecular species identification in canned tuna: Insights into human food safety. *Molecular Medicine Reports*, *15*(5), 3430-3437.
41. Pollard, S., Boyer, R., Chapman, B., di Stefano, J., Archibald, T., Ponder, M. A., & Hideout, S. (2016). Identification of risky food safety practices at southwest Virginia farmers' markets. *Food Protection Trends*, *36*(3), 168-175.
42. Pressman, P., Naidu, A. S., & Clemens, R. (2020).COVID-19 and food safety: risk management and future considerations. *Nutrition Today*, *55*(3), 125-128.
43. Piggott, N. E., & Marsh, T. L. (2004). Does food safety information impact US meat demand? *American Journal of Agricultural Economics*, 154-174.
44. Qiao, M., Liu, Q., Yong, Y., Pardo, Y., Worobo, R., Liu, Z.,& Ma, M. (2018). Scalable and rechargeable antimicrobial coating for food safety

applications. *Journal of agricultural and food chemistry*, *66*(43), 11441-11450.

45. Rendueles, E., Omer, M. K., Alvseike, O., Alonso-Calleja, C., Capita, R., &Prieto, M. (2011). Microbiological food safety assessment of high hydrostatic pressure processing: A review. *LWT-Food Science and Technology*, *44*(5), 1251-1260.
46. Redmond, E. C., & Griffith, C. J. (2004).Consumer perceptions of food safety risk, control and responsibility. *Appetite*, *43*(3), 309-313.
47. Sargeant, J. M., Saint-Onge, J., Valcour, J., Thompson, A., Elgie, R., Snedeker, K., &Marcynuk, P. (2009).Quality of reporting in clinical trials of preharvest food safety interventions and associations with treatment effect. *Foodborne Pathogens and Disease*, *6*(8), 989-999.
48. Smith, J. S., &Pillai, S. (2004). Irradiation and food safety. *Food technology*, *58*(11), 48-55.
49. Sadiq, N. W., &Beauchemin, D. (2017).Simultaneous speciation analysis of arsenic, chromium, and selenium in the bioaccessible fraction for realistic risk assessment of food safety. *Analytical chemistry*, *89*(24), 13299-13304.
50. Sinha, R. K. (2009). The concept of sustainable agriculture: an issue of food safety & security for people, economic prosperity for the farmers and ecological security for the nations. *American-Eurasian Journal of Agricultural and Environmental Science*, *5*(S), 1-4.
51. Todt, O., Muñoz, E., González, M., Ponce, G., &Estévez, B. (2009). Consumer attitudes and the governance of food safety. *Public Understanding of Science*, *18*(1), 103-114.
52. Tirado, M. C., Clarke, R., Jaykus, L. A., McQuatters-Gollop, A., & Frank, J. M. (2010). Climate change and food safety: A review. *Food Research International*, *43*(7), 1745-1765.
53. Wahyuni, H., Vanany, I., &Ciptomulyono, U. (2019). Food safety and halal food in the supply chain: Review and bibliometric analysis.*Journal of industrial engineering and management*, *12*(2), 373-391.
54. Wertheim-Heck, S., Raneri, J. E., &Oosterveer, P. (2019). Food safety and nutrition for low-income urbanites: exploring a social justice dilemma in consumption policy. *Environment and Urbanization*, *31*(2), 397-420.

55. Wilcock, A., Pun, M., Khanona, J., &Aung, M. (2004). Consumer attitudes, knowledge and behaviour: a review of food safety issues. *Trends in food science & technology*, *15*(2), 56-66.
56. Xiang, L., Li, Y. W., Yu, P. F., Feng, N. X., Zhao, H. M., Li, H., ... & Li, Q. X. (2019). Food safety concerns: crop breeding as a potential strategy to address issues associated with the recently lowered reference doses for perfluorooctanoic acid and perfluorooctanesulfonate. *Journal of agricultural and food chemistry*, *68*(1), 48-58.
57. Zanin, L. M., da Cunha, D. T., de Rosso, V. V., Capriles, V. D., &Stedefeldt, E. (2017). Knowledge, attitudes and practices of food handlers in food safety: An integrative review. *Food Research International*, *100*, 53-62.

INDEX

Printed in the United States
by Baker & Taylor Publisher Services